SAR In
Interpretation for
Various Land Covers
A Practical Guide

SAR Image Interpretation for Various Land Covers

A Practical Guide

Élizabeth L. Simms

CRC Press
Taylor & Francis Group
Boca Raton London New York

CRC Press is an imprint of the
Taylor & Francis Group, an **informa** business

CRC Press
Taylor & Francis Group
6000 Broken Sound Parkway NW, Suite 300
Boca Raton, FL 33487-2742

First issued in paperback 2022

© 2020 by Taylor & Francis Group, LLC
CRC Press is an imprint of Taylor & Francis Group, an Informa business

No claim to original U.S. Government works

ISBN 13: 978-1-03-247493-9 (pbk)
ISBN 13: 978-0-367-20996-4 (hbk)
ISBN 13: 978-0-429-26477-1 (ebk)

DOI: 10.1201/9780429264771

Visit the Taylor & Francis Web site at
http://www.taylorandfrancis.com

and the CRC Press Web site at
http://www.crcpress.com

For my husband, Alvin

Contents

Preface

Introductory remote sensing courses address several topics, the entire electromagnetic spectrum, and standard analysis processes. This gives a learner the opportunity to become familiar with the technology and benefit from applications in the context of their work. People intuitively use geographical information to find places of interest. The visualization of images of the Earth is routine procedure for on-line interactive geographical location searches, travel-path finding, or weather-condition seeking.

As an instructor of remote sensing, limited time is available to divert from an established curriculum that is developed to serve most of the students' interests, learning goals, and professional development. As a researcher, I have been captivated by the visual information that synthetic aperture radar images contain and the possibilities they offer to learn about our environment.

This book presents a collection of images that portrays land use and land cover related topics and synthetic aperture radar-specific remote sensing concepts. This encompasses passions that I have developed over more than thirty years of teaching and research about gathering information to describe landscapes and help understand geographical processes.

The purpose of this book is to demonstrate the utilization of interpretation keys for land use and land cover information extraction from synthetic aperture radar images. The work contains a summary review of image acquisition parameters of consequence on the visual representation of objects, as well as introduces traditional interpretation keys under a different light and applies them for considering regional landscape components and identifying large-scale geographical ensembles.

The first version of this book was a website that proved to be difficult to maneuver. The format gravitated toward a simplified image-set, a few broad topics, and labeled technical-like sheets. The current publication is a compromise to the exploratory versions. It includes several images, focused topics, two spatial scales of representations, and numerous coordinate-located explanations. The label-free images can be used to study the concepts, investigate user-defined examples, or discover other geographical ensembles and objects.

The project is based on synthetic aperture radar images from Cosmo-SkyMed and Radarsat-2 acquired for seven locations in Canada and Italy. The areas of interest were selected for the opportunities to compile images for a variety of land cover types, with different seasons and environments for a backdrop. The period of observation spans from 2008 to 2015, which represents the time I spent to conceptualize the book, adjust the format, and prepare the material. With the new Cosmo-SkyMed and Radarsat missions continuing to exploit the X- and C-band, respectively, an inexhaustible source of data will be available for applying notions that can be learned through this book.

Interpreting images acquired using non-visible spectral bands draws on one's open-mindedness, but learning from what could be described as an abstraction is very rewarding. Captured using a technology unfamiliar to most, the information that synthetic aperture radar images hold is revealed to be greatly accessible. The book is unique because of the large number of illustrations it contains. The chapters and sub-chapters discuss the geographical topics, and each figure is accompanied by a caption that describes the features and points to their coordinates within the image frame. Ancillary information includes acquisition specifications, a geographic scale, and the image-center latitude and longitude.

Overall, the book can be approached as a visual geographic lexicon. Readers, and also onlookers, are invited to linger for making observations, creating interpretation hypotheses, and building a visual memory of how the surface of the Earth is represented on synthetic aperture radar images.

Acknowledgments

Radar images for this project were provided by the Italian Space Agency and the Canadian Space Agency. It was a privilege to have my project approved and permission granted to reprint in a format for publication the images acquired by COSMO and Radarsat-2 synthetic aperture radar systems. The agencies sponsored utilization of images for the Science and Operational Applications Research (SOAR) project 'Cosmo-SkyMed and Radarsat-2 for land cover interpretation' (Project numbers SOAR 5252 and SOAR-ASI LI-11525-21).

The COSMO-SkyMed initial product was provided by the Italian Space Agency under copyright agreement 'COSMO-SkyMed Product - ©ASI - Agenzia Spaziale Italiana (2009, 2010, 2013, 2014, and 2015). All Rights Reserved'. RADARSAT-2 initial products were obtained through the project name 'COSMO-SkyMed and RADARSAT-2 for Land Cover Interpretation' and number 5252 of the Canadian Space Agency. The images were made available under the copyright and credit to 'RADARSAT-2 Data and Products ©MacDonald, Dettwiler and Associates Ltd. (2008, 2014) - All Rights Reserved. RADARSAT is an official trademark of the Canadian Space Agency.'

I acknowledge the dedicated and meticulous work of students hired through the Memorial University Career Experience Program at the beginning of the project. Michael Upshall and Sarah TeBogt investigated potential locations for image acquisition, prepared image subsets, and compiled geographic information.

I am grateful to Ms. Anne Bown and Mr. Harry Bown for their availability to make a review of my manuscript. It was considerably improved by their comments.

Many thanks to Mr. Glenn Crewe for his valuable contribution of reviewing and discussing my manuscript.

I also thank my family and friends for their heart-warming enthusiasm about my book-writing project.

Biography

Élizabeth L. Simms is a professor in the Department of Geography at Memorial University. She received her MSc degree from the Université de Sherbrooke and completed her Ph.D. from the Université de Montréal. She worked for the Canada Centre for Remote Sensing and C-CORE on research projects related to the ocean, coastal environment, agriculture, and natural resource monitoring. She joined Memorial University in 1990. Her academic activities include teaching courses in remote sensing, introductory geographic information sciences, field methods, research design, and quantitative methods. Administrative positions Dr. Simms held at Memorial University consist of the Diploma in Geographic Information Sciences program coordinator, Head of the Department of Geography, and Associate Dean of Research and Graduate Programs, Faculty of Humanities and Social Sciences. Dr. Simms supervised graduate students in research applying synthetic aperture radar and multispectral images for the analysis of boundary environments such as the coastal zone, Arctic tree line, and glacier ice margins. Other research interests include developing teaching material to assist with learning based on Radarsat-2 and Cosmo-SkyMed images, assessment of developed land cover type and intensity from synthetic aperture radar, court case applications of remote sensing, and the evaluation of remote sensing classification schemes for the representation of landscape features described through Aboriginal language.

List of Abbreviations

CORINE	Coordination of information on the environment
CSK	Cosmo-SkyMed
EHN	Enhanced Spotlight image mode of Cosmo-SkyMed
FQn	Fine-Quad image mode and Incidence angle position (n) of Radarsat-2
HH	Horizontal send and Horizontal receive polarization
HIMG	Stripmap HImage image mode of Cosmo-SkyMed
HV	Horizontal send and Vertical receive polarization
ID	Nominal image identification number
LA	Left looking from Ascending orbital path
LD	Left looking from Descending orbital path
LL	Latitude and Longitude
LULC	Land Use and Land Cover
PCA	Principal Component Analysis
RA	Right looking from Ascending orbital path
RD	Right looking from Descending orbital path
rms	root-mean-square height
RS2	Radarsat-2
Sn	Standard image mode and Incidence angle position (n) of Radarsat-2
SQn	Standard-Quad image mode and Incidence angle position (n) of Radarsat-2
Un	Ultrafine image mode and Incidence angle position (n) of Radarsat-2
USGS	United States Geological Survey
VH	Vertical send and Horizontal receive polarization
VV	Vertical send and Vertical receive polarization
W	Ground coverage Width

1

Introduction

Visualization of the World in different ways is enabling one to understand it better. New technologies and related applications have led to the establishment of an extensive Earth observation data consumer base. Numerous types of sensors and platforms address a need for high revisit frequency, near real time viewing, access to remote locations, and always a high spatial resolution and seamless geographical coverage. Increasing demand for remote sensing images is an important motivation behind the development of new systems and advancement of the potential that current technologies have to offer. In this context, further awareness of the information that spatial data conveys expands the opportunities to learn about the built and natural landscapes, and our environment as a whole. Publicly available remote sensing image products currently reach a broad audience and support daily usage of geographical information. The preferred representation is intuitive, as it is a real color composite based on the visible spectral bands. However, only a very small portion of the electromagnetic spectrum contributes to this dataset. Remote sensing research and development professionals tirelessly seek out the potential of different data types. Synthetic aperture radar images, which utilize the microwave spectral band, are of great interest because they provide high spatial resolution images and an unobstructed view of the Earth surface during most weather conditions. The atmospheric window of the microwave spectral band makes it an ideal candidate for land use and land cover (LULC) monitoring. Also, the synthetic aperture radar image product is operational, but a challenge to its use expanding resides in an endeavor for people to apply this technology. It entails that information-rich representations are made available to the public, who, further, need to see that the images are useful to them. To understand the data visually is an important and first step in accessing the information that it provides and in eventually considering relevant numerical analysis.

The purpose of this book is to guide the interpretation of land cover from synthetic aperture radar images. Its aim is to assist practitioners, researchers, students, and the public whose activities involve collecting geographical information. A wide variety of LULC topics are presented for applications in the context of readers' own tasks. The initial chapters explain some of the synthetic aperture radar technological specifications and establish the image interpretation approach. The chapters that follow describe the LULC examples at regional and large geographical scales. Throughout, an account of the literature on relevant synthetic aperture radar applications is given.

The chapter, Synthetic Aperture Radar Images: An Overview, describes technological designs and options of the synthetic aperture radar system that are expressed through the visual appearance of objects and surfaces. Consequently relevant to the image representation are the incidence angle, orbital path and look direction, polarization, spatial resolution, and wavelength. Concepts are derived pertaining to relative placements within a synthetic aperture radar image display. These include the azimuth and range orientations, near and far range, and terrain slope with respect to the incident radar beam.

The chapter, Image Interpretation Keys, reintroduces elements that were initially developed for LULC classification from aerial photographs and contextualize their application with synthetic aperture radar images. The keys are image tone, texture, pattern, shadow, shape, and dimensions. The systematic consideration and convergence of clues according to the interpretation elements will help to form descriptions.

The chapter, Regional Land Cover Descriptions, presents the interpretation carried out from full synthetic aperture radar scenes. The different sections expand on information about the location and discuss the visual characteristics of a dominant land cover type. The reader will become familiar with the main images, from which LULC examples were selected. Regional geography background supports the interpretation, with emphasis on a coastal zone, forest lands, river valleys, mountainous regions, agriculture, and a northern location.

The chapter, Large Scale Geographical Ensembles, displays annotated synthetic aperture radar image examples for thematic categories that inform about the LULC, geographical ensembles of constructed and natural elements. The illustrations were taken from images acquired for the locations introduced in the chapter Regional Land Cover Descriptions.

The Image Acquisition Specifications into Effect chapter presents an overview of the benefits offered by different image acquisition specifications. Applications of incidence angle options, opposite look directions, polarization modes and spatial resolution point to selections that create information accessible through visual image interpretation. In addition to reporting on selected research outcomes, this chapter includes a comparative presentation of images obtained with different acquisition specifications.

The book concludes with a discussion of synthetic aperture radar image products as a means to extract information about the LULC and to provide an up-to-date visualization of the state Earth surface. A list of references and an index are included. Cosmo-SkyMed (CSK) and Radarsat-2 (RS2) images illustrate different concepts, land use, land cover, and geographical ensembles. These were identified with the help of ©2018 Google images, Street View, and Historical Imagery applications of Google Earth Pro. Several interpretation solutions were substantiated with publicly available documentation and geography background literature. Each image is accompanied by an explanatory caption and bordered by a ten-grade-bottom bar and a five-unit-left-side bar (Figure 1.1).

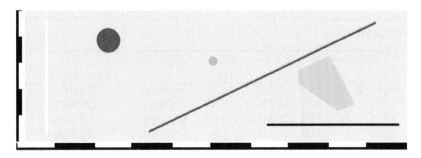

FIGURE 1.1
The image includes six objects on gray tone background. Three lines are oriented differently: a black vertical (1-right,1-5), blue diagonal [(4,1) to (10,5)], and white horizontal (7-10,1). Distributed throughout from left to right are blue circle (3,4-5), a green point (5-left,4-low), and a yellow pentagon (8-9,2-3).

These establish the visual reference for plane Cartesian coordinates used in locating objects to which the captions refer. Thus, coordinate statements use the (x,y) notation format, where 'x', i.e., horizontal axis, takes a value of 1 to 10 and 'y', i.e., vertical axis, ranges from 1 to 5. In some cases, where needed, a ten-unit y-axis and further locational details are given by mentioning a 'right' or 'left' placement within the horizontal units, and 'low' or 'high' in the vertical units. The image caption closing statements include the synthetic aperture radar system name, i.e., Cosmo-SkyMed or Radarsat-2, a three digit identification number that refers to further information about the image (Table 1.1), acquisition specifications, and geographical location. Parameters listed are applicable to some interpretation tasks. Of interest are the polarization option, incidence angle, orbital path and look direction, and the main image place name, the subset ground width (W), and area center Latitude Longitude (LL) coordinates.

TABLE 1.1
List of Cosmo-SkyMed (CSK) and Radarsat-2 (RS2) images used in the book, with an identification number (ID), the beam mode, nominal spatial resolution (NSR), main place name and acquisition date. Further specifications are provided with each image subset throughout the book, in their respective caption closing statement

Satellite	ID	Mode	NSR (m)	Place name	Date (yyyy-mm-dd)
CSK	006	HIMG	5	Ferrara	2014-05-14
	032	HIMG	5	Ottawa	2014-08-24
	034	HIMG	5	Ottawa	2014-08-16
	036	HIMG	5	Ottawa	2014-08-08
	055	HIMG	5	Taranto	2014-06-29
	208	HIMG	5	Labrador	2013-12-09
	212	HIMG	5	Cortona	2014-06-23
	433	HIMG	5	Montréal	2014-07-20
	806	HIMG	5	Ferrara	2014-05-13
	840	HIMG	5	Taranto	2014-06-27
	906	HIMG	5	Fredericton	2015-06-25
	908	HIMG	5	Cortona	2015-07-08
	909	HIMG	5	Ottawa	2015-08-15
	925	HIMG	5	Labrador	2015-06-25
	936	EHN	1	Fredericton	2009-04-27
	953	HIMG	5	Fredericton	2009-03-26
RS2	086	SQ19	21	Fredericton	2008-07-30
	087	FQ8	11	Fredericton	2008-07-04
	088	U18	3	Montréal	2008-07-07
	474	FQ17	9	Labrador	2008-07-22
	475	SQ5	22	Labrador	2008-07-25
	476	U6	3	Montréal	2008-07-17

2

Synthetic Aperture Radar: An Overview

The wavelength and active remote sensing model from which synthetic aperture radar images are acquired offer unequaled opportunities to detect and identify entities of the Earth surface. Synthetic aperture radar systems exploit a portion of the electromagnetic spectrum that is beyond those with a tactile significance, notably the visible and thermal bands, to which humans can relate. Instead, the millimeter and centimeter scale wavelengths are of the same order of magnitude as surfaces or volumes found in natural and built environments.

A typical synthetic aperture radar image holds only one spectral band. Access to different polarization modes provides a rich multidimensionality, as well as the datasets that draw from the incidence angle and look direction programmability during descending and ascending orbital paths. However, an image acquisition schedule may include a limited or an inconsistent array of options. This book grants single-band representations, which simplifies the interpretation process and helps to relate visual information to image acquisition specifications.

This chapter outlines some system designs that have an effect on the appearance of objects represented on synthetic aperture radar images. Topics include the incidence angle, look direction, polarization, spatial resolution, and wavelength. These specifications ensue from the data available for this publication. Each section briefly describes an image mode or technological specification and an account of those from which the LULC interpretation examples were developed.

Cosmo-SkyMed and Radarsat-2 satellites carry synthetic aperture radar as main Earth observation systems. Their path follows sun-synchronous orbits. Ascending and descending trajectory segments give opportunities to look to the ground from different directions and possibly on either side, right or left of the orbital track. The radar transmitter and receiver are programmed for the beam to expose the Earth surface and detect backscattered energy over designated incidence angles and polarization options.

The Constellation of Small Satellites for Mediterranean basin Observation, Cosmo-SkyMed, includes four X-band synthetic aperture radar systems synchronized to carry out a geographic location revisit every few hours. Orbiting in solo, Radarsat-2 has a 24-day repeat cycle which may be upgraded using right and left look directions. The Italian Space Agency's Cosmo-SkyMed satellites were placed in orbit one at a time; two in 2007, then in each 2008 and 2010. This fleet will be renewed by Cosmo-SkyMed Second Generation Constellation and is composed of two satellites. Their launch is scheduled for 2019 and 2020.

Radarsat-2, launched in 2007 with the Canadian Space Agency as the main sponsor, replaced Radarsat-1 and precedes the three-satellite Radarsat Constellation Mission. Continuing from the previous missions, the Constellation utilizes C-band energy. Launched in 2019, the new satellites were simultaneously deployed in orbit to offer daily revisit for 90 % of the Earth's surface.

The two missions have similar lifespans and complementary acquisition specifications. They both operate with a relatively short wavelength of the electromagnetic radiation microwave region and, with their high temporal resolution, the newest constellations form a broad set of incidence angles, spatial resolution modes, and polarization options.

The interpretation examples included in the book are geometrically referenced to the Universal Transverse Mercator system following an even value in azimuth and range dimensions and the product nominal spatial resolution (Table 1.1). Unless otherwise indicated, all images displayed are oriented with North to the top. The Cosmo-SkyMed images were provided as amplitude values corrected for speckle noise. The downloaded Radarsat-2 Single Look Complex images were calibrated to 16-bit sigma backscattering values. The displayed image tones, proportionally, represent the natural logarithm transformation of these values, linearly rescaled to a positive dynamic range.

Prior to visual interpretation, in a few instances, digital analyses have been applied. Low-pass filtering generalized the images for Chapter 4 regional representations, so they fit the published page width. Chapter 6 includes comparative interpretations based on merged image sets. A simple multiplicative operation summarizes each of the like- and cross-polarization image pairs for them to be compared. The first component of four polarization sets, created through a non-standardized principal component analysis, helped assess the information from different spatial resolutions images. Selective low-pass filtering was applied in cases of image noise which impeded the interpretation of large scale geographical ensembles presented in Chapter 5.

2.1   Incidence Angle

Synthetic aperture radar images are acquired from an oblique view with the antenna pointing downward from the satellite platform. Incidence angle small and large values describe low-oblique to high-oblique radar beam vantages, respectively. Selectable incidence angles typically range from about 20 to 60 degrees. A small angle is closer to the vertical view that is best known to be used for creating visible band panchromatic or multispectral images. A large incidence angle is obtained by steering the radar beam away from the satellite orbital path. The incidence angle related beam mode is achieved by programming to take effect during the acquisition of a scene over a particular location. Likewise, an incidence angle array exists from the near to far range as width-wise placement within an image. The nearest range is close to the orbital path and has the smallest incidence angle within the image, while the farthest range is at the largest incidence angle. The value represents the angular separation between the emitted radar beam path and vertical where it touches level ground.

A horizontal surface's normal coincides with the vertical; on a sloped terrain, the normal differs from the vertical, therefore locally modifying the incidence angle. Abrupt slopes and vertical walls directly facing the radar beam are examples of configurations that alter the incidence angle. Visual aberrations formed in such conditions result from the foreshortening and layover of the image. On a radar beam exposed slope, the local incidence angle is smaller by as much as the terrain inclination value than the system incidence angle. At the location where a slope angle becomes equal to the beam incidence angle, foreshortening occurs. Where the radar beam is in line with the surface normal and the local incidence angle is zero a specular backscattering towards the radar causes a bright return named a specular point (Raney 1998). For a terrain slope that is even more abrupt, image layover appears. In this case, a far range portion of the radar beam forms the image of a surface at the location which should be on the near, just adjacent, range. Layover gives the appearance that tall objects, such as buildings and cliffs, are displaced in the radar antenna direction. High incidence beam angles are pre-disposed to generating foreshortening and layover. However detrimental to the true representation and location of objects, these image distortions are often helpful for the interpretation of urban land cover and landforms.

The incidence angle statement for a particular image is made in one of two ways. The first is to give the angle value at near and far range image extremities and the second is to use the nominal value at the middle image-width location. These are obtained from the remote sensing system technical sheets. The images displayed in this book apply almost the entire range offered by the Cosmo-SkyMed and Radarsat-2 systems, with middle incidence angles from 18.1 to 43.6 deg.

2.2   Orbital Path and Look Direction

While a synthetic aperture radar antenna emits energy according to a prescribed incidence angle, it points perpendicularly to the satellite orbital track. The look direction is this sideway alignment of the radar beam. A right or left look direction can be used in combination with the descending or ascending path. The near-polar orbit has a inclination from the Equator of 97.86

deg with Cosmo-SkyMed, and 98.60 deg with Radarsat-2. Therefore, for both satellites, the right look from a descending orbit is oriented about eight degrees North of West and a left look views toward eight degrees South of East. Conversely, from the ascending orbit, the right and left looks are toward the East of Northeast and West of Southwest, respectively. The satellite orbital path, or its azimuthal direction, determines the sequence of image rows and the look direction establishes the order of objects from one column to the next, which spreads from the near range to the far range.

Effects of the look direction take shape where the radar beam faces inclined terrain or a vertical object. Surfaces exposed to the radar beam are more likely to increase the backscattering intensity either because of a narrowing local incidence angle or foreshortening occurring. A surface that is turned away in the opposite direction increases the incidence angle locally or creates an obstacle. A radar shadow is formed on the image as the result of a radar beam portion not reaching the ground on the unexposed side of an object where the terrain slope is relatively abrupt or concave. Consideration of the look direction helps with the identification of buildings, trees, fences, vegetation patch edges, valleys, mountain ridges, and terraces. Object heights of even only a few meters may cause radar shadow. Bright and dark tones that represent the beam exposed and opposed sides, respectively, are particularly well contrasted for linear features that extend perpendicular to the look direction (Xia and Henderson 1997). In addition, interpreting is facilitated for linearly shaped specular backscatters, say of a road or river, to which the look direction is parallel (Ban and Jacob 2013). In this case, the backscattering is unobstructed and the resulting tone is potentially better contrasted with surrounding LULC types.

Another look direction-dependent image characteristic is the cardinal effect. It often occurs in high and medium intensity development land cover where buildings are aligned perpendicular to the radar beam and cause specular backscattering toward the receiver (Raney 1998). As a result, the tone is very bright and a grid pattern usually is perceptible.

A programmable look direction is an important system feature because it potentially provides views of the landscape from different perspectives. Added value from this technological asset resides with the amalgamation of perspectives for identifying of LULC and objects of various terrain shapes and alignments (Ban and Jacob 2013; Ge, Gokon, and Meguro 2019). Because they impact on the visual interpretation of many objects and their spatial arrangement on the landscape, look direction specifications are recurrently provided with each image presented in the book. Three look angle and orbit combinations were used for acquiring the images. They are the right look from the ascending (RA) and descending (RD) orbits, and left look from descending (LD) orbit.

2.3 Polarization

Electromagnetic energy holds electric and magnetic fields that are at a right angle to each other, and both are normal to the beam direction of propagation. A grate system sets a horizontal (H) or vertical (V) orientation to the electric constituent of the emitted radar beam and also controls the way by which the backscattered energy is received at the sensor. The polarization function statement contains two letters. The first letter gives the emitted beam orientation and the second indicates how it is received. The options are HH and VV for like-polarization, and HV and VH for cross-polarization. Synthetic aperture radar systems designs offer single polarization or a combination of options for creating as many images. The beam polarization may be selected to obtain information related to the fine scale construct of objects. Angular configurations that squarely reflect the emitted beam produce a like-return detected by the HH or VV options. Slanted or varied layouts that create a surface roughness may depolarize the radar beam. In this case, cross-polarization have the receiver capture a reoriented portion of the backscattered energy. All Cosmo-SkyMed images exhibited in this book are HH polarization. The Radarsat-2 dataset contains HH, VH, and quad-polarization images. The latter were simultaneously acquired using all four options. Differences noticed on images that were acquired with like- and cross-polarization options are highlighted in Chapter 6.

2.4 Spatial Resolution

The spatial resolution sets the horizontal dimensions of a surface represented by an image pixel. A high resolution has each pixel correspond to a small area. In azimuthal direction, which is row wise in an image array, the synthetic aperture radar image resolution remains constant but it increases from its near to far range. This effect is proportionally more important for images acquired at low incidence angle. The spatial resolution affects the perception of shapes, textures and patterns. Image interpretation is facilitated by higher resolution images because they more clearly reveal the shape of objects. Until specific outlines are perceived, spatial variability of tones is described by the image texture. Thus, patterns can emerge for a particular LULC over increasing spatial resolution continuum.

This image acquisition parameter is often referred to as 'pixel spacing' or 'pixel size'. This is also representing a ground dimension and it is determined by the system's spatial sampling rate to capture the frequency content in the image (MacDonald Dettwiler Ltd 2018). In keeping with the other acquisition parameters of incidence angle such as look direction and polarization,

different synthetic aperture radar systems offer a number of spatial resolution
options, by which the image beam modes are named. The Cosmo-SkyMed
images presented in this book were acquired in ENHANCED SPOTLIGHT
or STRIPMAP HIMAGE mode with nominal spatial resolutions of 1 and 5 m,
respectively (Agenzia Spaziale Italiana (Eds.) 1987). The Radarsat-2 images
are of the Ultrafine, Fine or Standard mode which have nominal resolutions
of 2.5, 10, and 20 m, respectively (MacDonald Dettwiler Ltd 2018).

2.5   Wavelength

Synthetic aperture radar images represent energy from the microwave portion
of the electromagnetic spectrum, whose wavelength is of the same order as
surface height it exposes with which it interacts. Wavelength is the distance
between crests of the electromagnetic radiation as a wave. Wavelength and
incidence angle regulate the radar beam backscattering as specular, diffuse, or
combined fashion. Short wavelengths enable one to differentiate between specular and diffuse backscattering among millimeter scale surface irregularities.
Long wavelengths help with demarcating among coarser surfaces. Meanwhile,
larger incidence angles facilitate the detection of fine surface textures.

The Cosmo-SkyMed and Radarsat-2 images were acquired with a X- and
C-band which are 3.1 and 5.5 cm middle wavelength, respectively. Surface
roughness estimates based on the modified Rayleigh Criterion (Peake and
Oliver 1971; Xia and Henderson 1997) show that little difference separate
surface heights for fine, intermediate, and coarse textures with each image
type. The surface heights to categorize the texture parameter were calculated
given near and far range incidence angles of 18.1 and 41.0 deg, respectively, for
Cosmo-SkyMed (Table 2.1), and 19.4 and 43.6 deg for Radarsat-2 (Table 2.2).
Both these systems, with their centimeter scale sensitivity, provide similar
capabilities to differentiate specular from diffuse backscattering surfaces; they
convey specular backscatter related information where the surface height is
less than 0.4 mm and behave as diffuse backscatters where the roughness

TABLE 2.1
Root-mean-square (rms) height for fine, intermediate, and rough surface textures calculated from the modified Rayleigh Criterion based on the Cosmo-
SkyMed X-band (3.1 cm) synthetic aperture radar, and near and far range
incidence angles of 18.1 and 41.0 deg, respectively

Surface Texture	Fine	Intermediate	Rough
Near range rms height (mm)	<0.4	0.4 to 2.2	>2.2
Far range rms height (mm)	<0.8	0.8 to 4.6	>4.6

TABLE 2.2

Root-mean-square (rms) height for fine, intermediate, and rough surface textures calculated from the modified Rayleigh Criterion based on the Radarsat-2 C-band (5.5 cm) synthetic aperture radar, and near and far range incidence angles of 19.4 and 43.6 deg, respectively

Surface Texture	Fine	Intermediate	Rough
Near range rms height (mm)	<0.7	0.7 to 4.2	>4.2
Far range rms height (mm)	<1.5	1.5 to 8.6	>8.6

is above 8.6 mm. Intermediary values describe surface heights that produce mixed backscattering of specular and diffuse forms.

2.6   Summary

Cosmo-SkyMed and Radarsat-2 synthetic aperture radar images constitute the dataset from which examples are produced for this book. A brief overview of the satellite missions and their continuity is reported in this chapter. As newcomers to the remote sensing public realm, the synthetic aperture radar images are presented focusing on the way by which it is acquired influences the appearance of objects, land cover, land use or geographical ensembles. Technological specifications of incidence angle, orbital path and look direction, polarization, spatial resolution, and wavelength are each, in one or many aspects, important to keep in mind while visual clues are considered during the image interpretation process. These system design parameters are summarized, and references point to details documented in the literature, which offers rich theoretical background. The chapter also presents the terminology applied in the book preparation, a nomenclature consistent with current practices for synthetic aperture radar image acquisition planning or ordering.

3

Image Interpretation Keys

The set of specifications by which a radar beam is emitted and received is meant to maximize the information available to identify, characterize and measure landscape components. Because it is not hindered by most weather conditions this long wavelength imagery's temporal resolution is functional. Aerial photographs and early generation multispectral images relied on passive remote sensing systems, a vertical view, limited use of polarization, and short wavelength electromagnetic radiation. Conversely, synthetic aperture radars exploit the active remote sensing design by acquiring images from an oblique perspective with polarization options and long wavelength energy. These distinct remote sensing models prompted to assess the way by which visual keys, or elements, that initially were developed for aerial photographs interpretation can be applied with synthetic aperture radar images.

The identification of objects and geographic ensembles ensues from examining a number of elements. First to consider are the image tones, then their spatial organization defined in the realm of textures and patterns. This second hierarchic interpretation level enables a description of the tone variation, and their contrasts or lack thereof. Differently from methods previously suggested (Estes, Hajic, Tinney, et al. 1983), information from observing shadows, shapes and dimensions assists an exhaustive postulation of the object characteristics. Each interpretation key is explained in the context of synthetic aperture radar image technological specifications, and examples of their visual expression are paired with terms that describe their characteristics, e.g., bright, intermediate, dark tones. This chapter establishes the image interpretation methodology that is applied in the subsequent chapters for interpreting LULC types, their characteristics and state, at regional and large spatial scales.

3.1 Tone

On the synthetic aperture radar magnitude image, the tone represents energy power redirected to, and received by, the radar antenna after the transmitted beam will have exposed a surface on the ground. Images are digital products and the representation of an image is analog. Whether it is as a virtual computer monitor display or a tangible picture on paper, the analog product is compatible with the human visual system. A single band synthetic aperture radar image is suited to a monochrome representation that has the tone developed on a spectrum from dark to bright. The digital values may be quantified or calibrated according to a defined framework, but typically, the assignment of tones corresponds to the image digital values signifying the energy power received at sensor. A relative suite of tones must be established for visualizing the image. Tones are described by at least two categories, dark and bright. However, a binary scheme, which includes only two tones, limits the abilities to evolve further into interpreting an image using other keys. Three, five, or seven levels can address operational requirements and the interpreter's experience. An odd number of levels is convenient, as it allows for the inclusion of a middle category. The descriptions that will be applied is this book are: very dark, dark, intermediate, bright and very bright. According to one's abilities or the need for details, additional levels are available. A practical method for working through the observation of tones is to take dark, intermediate and bright tones for the largest proportion of the image area and then use the extreme tonal values, very dark and very bright, for the outliers, or infrequently occurring instances. Areas that display little or no tonal variation are more easily described than those of coarse image texture, for which case the predominant tone must be assessed (Figure 3.1). The tone is beneficial to image interpretation in at least two ways. It delivers information about the exposed surface of the ground, which causes the radar beam to return as diffused energy or to specularly-backscattered away from the antenna. Second, considering a comparative assessment, it helps understand the spatial distribution of objects. Therefore, any bright and dark tone sequence could reveal information about the landscape structure even if, overall, a limited number of tones are used to categorize LULC types.

The type of backscattering largely controls image tones. Once energy beams have been emitted, initial interactions with the surface upon which it is incident determines the type of backscattering. These may be diffuse or specular, and a continuing beam pathway may be subject to corner or double-bounce backscattering. Diffused energy results from the radar beam being changed to an incoherent signal of various polarization and propagation angles, some of which enable an energy return toward the receiving antenna. It is a type of backscattering that is likely to produce a bright tone on the image. Diffuse backscattering is caused by rough terrain surfaces, which minimum height is determined by the radar beam wavelength and incidence angle according to the modified Rayleigh Criterion (Peake and Oliver 1971). Specular backscattering occurs on the smoothest of surface textures. Their millimeter scale height causes a redirection of the energy beam in the direction opposite to the antenna, resulting in a dark tone on the image. Tone variability is produced by intermediate surface textures and volume backscattering. Radar beam reflections by a bulk material, such as snow, leaves, branches or loose soil, occur when the particle dimensions and inter-spacing are of the same order as, or larger than, that of intermediate or coarse surface texture. A corner reflection

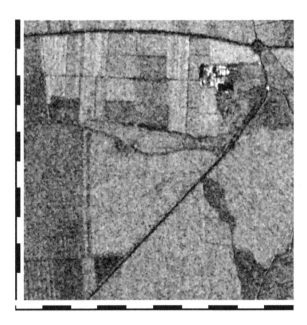

FIGURE 3.1
Agglomerated very bright and very dark tones (8,8-9) approximately define a building shape. Extreme values are also at road intersections [(9-right,9) & (10-left,8)], and mark a water body (5-6,7-low). Elsewhere in the image, fields and roads provide a range of tones that can be described in at least five categories. Darks typically outline roads (1-10,10) while several levels represent different agricultural crops' growth stage, type and condition. In this image, tones for the fields can be distinguished, in increasing order, from dark to bright at the locations [(4,2-low), (2-right,1), (1-2,4-6), (5-6,5), (2,9), & (9,9-high)]. Image: CSK 840 HH 31 deg RD Taranto W 1.5 km LL 40.444160 17.326664 deg.

submits the incident beam to at least two specular bounces, the first on a horizontal surface, and the second on a vertical facade. This results in a powerful signal return to the synthetic aperture radar receiver.

3.2 Texture

Image texture is a product of the spatial variation of tones. The observation of similarities and contrasts of tones, and their spatial frequency, inform this interpretation key. Fine and coarse image textures denote low and high spatial frequency tone changes, respectively. Visually, the texture can be described on a continuum from fine to coarse. With the use of an intermediate category, three levels are adequate for describing this interpretation key. Meanwhile, image texture and surface texture are concepts that must be set apart; the former is about the tone variation and it is in the image interpretation domain, and the latter concerns a ground surface that is exposed to the incident radar beam. Likeness of tones, of any value, produces a fine image texture, which indicates that a same backscattering behavior, specular or diffuse, occurs from place to place. Conversely, where variations happen in the manner by which objects in close vicinity to each other backscatter the radar beam, a coarse image texture emerges (Figure 3.2 and 3.3).

As discussed earlier, the tone is caused by that diffuse, specular or corner backscattering prevailing on the surface, or objects through a volume. Texture is formed by tonal differences of nearby pixels about each other. Therefore, any one image texture may be observed in association with an overall dark, intermediate or bright tone. For example, calm water appears dark and crop fields have bright tones and both typically display a fine image texture, because, while covering a reasonable size area, they uniformly backscatter microwave energy in specular and diffuse manner, respectively. Intermediate and coarse image textures are formed when the radar beam interacts with a

FIGURE 3.2
The texture gives insight into the spatial variability of tones, i.e., this image, by at least three levels. A fine texture applies to a dark tone water body (3,1-low) and a bright tone field (10,2). A coarse image texture pertains to an intermediate tone forest (3,5) and a bright tone for grasses mixed with piles of sediment (4,1). The intermediate texture level is subtle, especially that it consists of different tones. Some are found at the location of a landfill (3,2), and straddle a bright tone crop and dark tone treed grounds (6,3). Image: CSK 909 HH 18 deg RA Ottawa W 3.8 km LL 45.313166 -75.542048 deg.

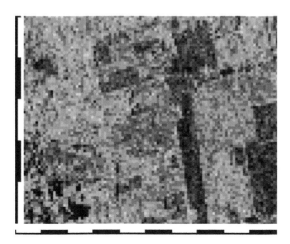

FIGURE 3.3
A digital texture image for the same area as Figure 3.2 readily shows fine to coarse textures with dark to bright tones, i.e., that represent similarities and contrasts, respectively, of backscattering-related tones. In this case, the texture image portrays the number of different brightness values in a 7-column by 7-row neighborhood of the backscattering-related image. Image: CSK 909 HH texture 18 deg RA Ottawa W 3.8 km LL 45.313166 -75.542048 deg.

surface patchiness or object that is at least in same order of size as the image spatial resolution. For example, tightly closed forests, large asphalted surfaces, lawns, and unconsolidated material tend to display fine textures, while open canopy forests, high-intensity residential developments, and fragmented land cover usually realize coarse image textures or patterns.

The texture is a particularly relevant interpretation key when the variation of tones is caused by objects that are difficult to distinguish from each other due to a relatively low spatial resolution. Whether it is fine or coarse, the image texture displays an overall evenness. By increasing the spatial resolution, the finest of image textures remain fine because they are a consequence of specular backscattering. However, the image textures that result from some diffuse backscattering may become coarser if the spatial resolution is increased, to the point where small surface patches and dispersed objects display a tone and texture of their own, among larger areas, that may exhibit identifiable patterns. A mix of radar beam foreshortening, shadows, and volume backscattering typically structures an intermediate or a coarse texture on synthetic aperture radar images. These three features are likely to appear in close proximity to each other for representing tall objects.

The images in the book are color composites that combine tone and texture (Figure 3.4). The tone is backscattering-derived and the texture is the number of different values in a 7-column and 7-row filter calculated from the fifty-percent-median group of brightness values. The synthetic aperture radar image value distribution functions are typically positively skewed. The infrequently-occurring very bright values that may represent, for example, point specular backscattering and corner backscattering are enhanced through the third color composite layer.

FIGURE 3.4

The color composite blue, green, and red layers represent the texture (Figure 3.3), backscattering-derived tone (Figure 3.2), and bright-tone outliers, respectively. Dark colors are the expression of specular backscattering and fine texture such as that of calm water surfaces (3,1) and crops [(5,4-low) & (7,4)]. Green-dominated colors indicate strong diffuse backscattering (i.e., bright tone) and a very-fine to fine image texture, for example, large crops with evenly growing vegetation [(10,2) & (4,5)]. Cyan mixed with green represents intermediate tone and coarse texture for some agricultural fields [(9,3) & (4-5,1)]. Dark blue to cyan composed color indicate spatially uneven dark or intermediate tones and a coarse texture. This can be caused by unevenly shaped and distributed objects onto which specular and diffuse backscattering occur from place to place. It is the case for tree canopies that may vary in height, species, or density [(3,4) & (10,5)]. High spatial frequencies such as field boundaries (7,1-3) and isolated objects [(6,4) & (8,4)] are enhanced by the relatively bright tone and coarse texture that yield a bright cyan color. The brightest of backscattering-related tones that occur with a fine texture context are in yellow [(7,3) & (1,2)]. A red color that emerges next to yellow (2, 2-low) indicates that bright-tone outliers are inside a coarse texture area. On other color composite representations of the book, white and pink also highlight bright tones in high spatial frequency variation neighborhoods. Image: CSK 909 HH 18 deg RA Ottawa W 3.8 km LL 45.313166 -75.542048 deg.

3.3 Pattern

The pattern expresses a repetitive sequence of tone changes. This interpretation key is applicable where the spatial variation of tones is caused by objects that can be visually separated from each other, delineated, and potentially identified. Comparatively, a texture is formed by objects whose dimensions are less or about equal to the image spatial resolution. For example, where ground is covered with permanent crops and trees grow in one or two meter apart rows, a high spatial resolution image may lend a dotted pattern where each dot represents a tree (Figure 3.5). Low spatial resolution image of a similar type of crops would display a coarse texture from a mix of ground and trees contrasted backscattering, and radar shadow. The terminology for describing a pattern may evoke familiar forms such as dotted, grid, mottled (Figure 3.6), patchwork (Figure 3.7), ribbed, speckled (Figure 3.8), striped (Figure 3.9), or tiled. The more contrasted image tones are the easier is the pattern assessment. Whether a specific description is proposed as the interpretation key of land cover for a particular area, it is useful to consider overall patterns or spatial arrangement of several side-by-side entities. To observe that an ensem-

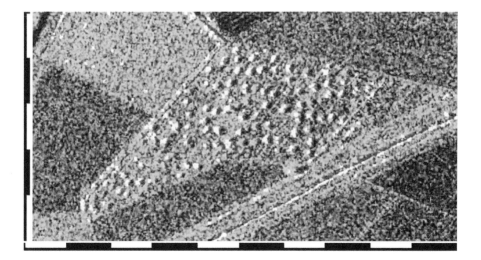

FIGURE 3.5
Dotted irregular patterns are formed by contrasted bright tone tree crowns and dark radar shadow on intermediate (4,3) and dark (7,3) tone field backgrounds. Surrounding planted herbaceous lands display a faint striped pattern (2,3-5) or homogeneous, not patterned, intermediate image texture (8-10,5). Image: CSK 006 HH 39 deg LD Ferrara W 1.5 km LL 44.581868 11.734706 deg.

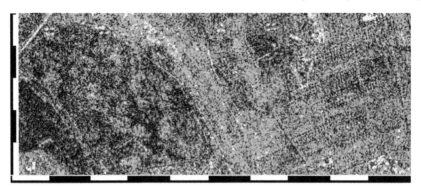

FIGURE 3.6

A mottled pattern is defined as large, more or less rounded, and irregularly distributed, shapes and spots (3-5,1-4). Little tone changes typically characterize a mottled pattern. In the case of sharp contrast, homogeneous areas can be delineated and characterized. As a low spatial frequency variation, mottling may be the background to another pattern, such as a fine regular dotted in the image eastern portion, (9,2-4) or stripes. Groups of bright dots for aligned pairs of electric power transportation towers (2-7,5) and clustered buildings (8,4-5) are noticeable by their very bright tones. Radar shadows also mark some field edges (9,3). Small objects, even if their backscattering is very different, often have attenuated contrasts on a mottled pattern. Image: CSK 840 HH 31 deg RD Taranto W 2.3 km LL 40.474102 17.366917 deg.

FIGURE 3.7

Patchwork is created from a mix of large, small, and regular or irregular geometrical shapes. Where units are evenly sized and distributed, the pattern may be described as tiled. In this example, from a variety of alternating contrasted tones and dimensions emerges an irregular patchwork. Individual and groups of fields are in places outlined by bright tone fences, ditches, or roads (8-left,2-5). Groups of buildings are dispersed throughout the area; most of them display bright tones [(8-left,3-high) & (9-left,2)]. Image: CSK 006 HH 39 deg LD Ferrara W 2.7 km LL 44.597731 11.621718 deg.

FIGURE 3.8

A speckled pattern is realized by very contrasted tone randomly distributed small dots, which are made of a few or groups of pixels. The image displays three land cover types distinct by their overall homogeneous texture and pattern. A herbaceous land parcel (1-2,4-5) and forests [(1-5,1-2) & (10,5)] are intermediate and coarse in texture, respectively, and a low-intensity residential development established among a heavily tree-covered landscape (7-9,1-4) prominently displays a speckled pattern. Image: CSK 055 HH 29 deg RA Taranto W 2.8 km LL 40.471664 16.928447 deg.

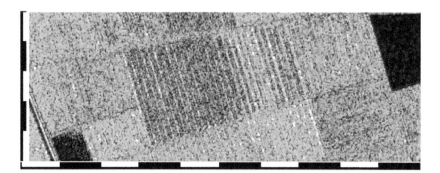

FIGURE 3.9

A striped pattern is formed of repetitive lines or bands. A regular spacing and straight shapes often indicate human imprint on the landscape. This pattern may emerge at different spatial scales. For example, an agriculture land use that involves allotment, plowed and planted rows, and water management infrastructure. Bright lines appear continuous across fields on a background of multi-tone tiles that indicate variety of crop type and growth stage. Image: CSK 006 HH 39 deg LD Ferrara W 2.8 km LL 44.685186 12.030784 deg.

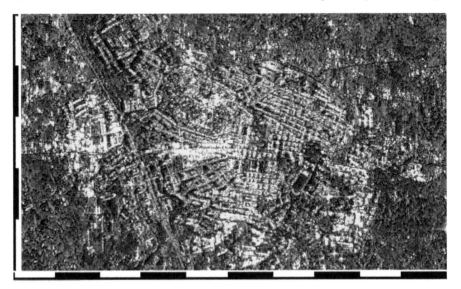

FIGURE 3.10
A grid pattern is omnipresent through city landscapes. Bright and intermediate tones represent buildings alternating with very dark radar shadows or
intermediate tone grounds. In blocky sections, the pattern is regular (7,3-4)
or irregular (5,2). In the center, likely the oldest city section (5-left,4), buildings are so close to each other that a coarse texture is formed, rather than a
pattern. Bright specks representing low-intensity developments are dispersed
in surrounding lands to the west (1,3-5) and clustered along roads (9-10,4-5).
Image: CSK 840 HH 31 deg RD Taranto W 3.5 km LL 40.702280 17.337954
deg.

ble is presenting a regular or irregular pattern (Figure 3.10) does provide
information, for example, about terrain topography constraints, hydrology
and other environmental parameters, or anthropological design practices. As
an overarching description, patterns may evoke the spatial distribution such
as dispersed, random, or clustered. These generic categories, that can be visually and quantitatively assessed (McGrew, Lembo, and Monroe 2014), convey
LULC information of interest in geographical studies (Simms 2017).

3.4   Shadow

A radar shadow appears as a very dark tone. It marks the far range side of an
abrupt slope that prevents the beam from reaching the surface at this location.

Occurrence of shadow is determined by the angular relationship between the radar beam incidence and the terrain slope value that is turned away from it. If the complementary angle to the beam incidence angle, which is the depression angle, is smaller than the terrain slope, the condition exists for radar shadow to be formed (Lewis and Henderson 1998). It extends along the image range direction up to where the image represents a surface receiving the radar beam that is either a gentler lee slope, a plane surface, or a beam exposed slope. Often by the radar shadows, on the image near range side, bright tones result from a direct exposure on a vertical or abrupt slope.

Shadows are very useful to interpret the terrain layout with respect to objects such as buildings, trees, fences, mountain ridges, valleys, and other symmetrical topographies (Figure 3.11) along the radar beam range path. The relative height of objects is an additional piece of information that can be

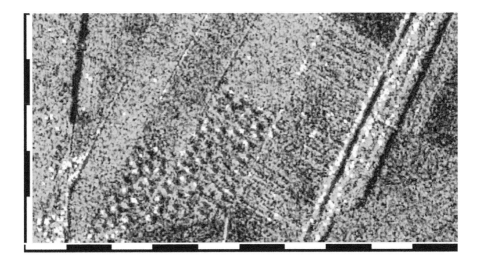

FIGURE 3.11
Interpreting characteristics of objects from radar shadows must consider the satellite orbital path and beam look direction. For instance, this image was acquired looking to the left from a descending orbit, noted as LD in the image specifications. Where tree crops are located (5,3), a dot pattern is formed of bright dot and radar shadow, in pairs, on near and far range sides, respectively. Azimuth-oriented linear objects are outlined by variable thickness shadows. Thicker ones paired with bright linear coarse textures are tree rows [(7,1) to (9,4), & (9,2) to (10,5)]. A single dark thick line is a shadow associated to the edge of a forest canopy (2,3-5). A thin and uneven shadow enhances shrubs and sparse trees along a road and by a farm house (2-left,1-2). Image: CSK 006 HH 39 deg LD Ferrara W 1.5 km LL 44.584326 11.714068 deg.

TABLE 3.1
Shadow lengths (SL) for different height vertical objects at near and far range locations with an incidence angle of 20 and 40 deg, respectively

Height (m)	Object Example	SL_{near}(m)	SL_{far}(m)
2	Tree	0.7	1.7
3	House	1.1	2.5
20	6-Story building	7.3	16.8
60	20-Story building	21.8	50.4

obtained from assessing radar shadow lengths. For this purpose, however, one must consider that throughout an image, equal height objects are represented by radar shadows that increase in length from the near to far image range, due to the incidence angle becoming larger in that direction. Shadow lengths are modest visually, but important to the interpretation process (Table 3.1). Acquisition specifications can be chosen to control the formation of radar shadows. A small incidence angle minimizes their occurrence and, conversely, a large angle is more likely to form distinct shadows. Information gaps are created by a constant perspective that systematically captures portions of mountainous areas or slope sides as radar shadow.

Images acquired from different look directions are necessary to fill-in backscattering related information for these areas. Complementary geographical coverage can be obtained from opposite look directions, for example descending-right and -left, or offset azimuthal directions, ascending- and descending-right.

3.5   Shape and Dimensions

Object shapes outlined from radar images may be generic or specific. Assessing a shape requires the recognition of a coherently illustrated single feature or an area that is relatively homogeneous by its tone, texture or pattern. This area may have determined minimum dimensions, traditionally known as the minimum mapping unit (Lillesand and Kiefer 1994), and perhaps associates with existing ecological or management zones (Sahebi, Bonn, and Bénié 2004). The description of shapes refers to standard geometrical configurations, such as a square, circle, oval, triangle, or rectangle area, line, or point (Figure 3.12). Ensembles' arrangement may evoke specific descriptions such as cloverleaf, crescent, u- or v- shapes. Meanwhile, extreme backscattering levels may have a star-shape feature formed on synthetic aperture radar images (Figure 3.13).

Tied to the shape, relative dimensions and spatially associated objects are also important to the land use and cover identification. When they are

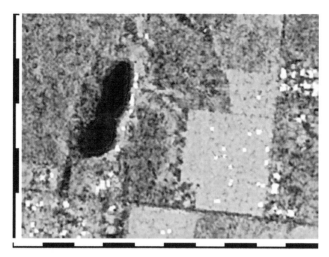

FIGURE 3.12
In this example, point-like bright and dark features include buildings (9-right,2) and small-area open lands (4,1), respectively. Linear features include roads [(1,2-low) to (10,1) & (9,1-5)], field boundaries [(7-left,1) & (8,5)], and vegetation-outlined stream paths [(5,5) to (6,2), & (6-7,5)]. Area features have different shapes that represent forests (1-2,4), a lake (4,4), farm fields (7-8,2-5), and developed lands [(4-left,2-low) & (10,4)]. Image: CSK 909 HH 18 deg RA Ottawa W 1.2 km LL 45.469887 -75.883722 deg.

on level terrain or moderate topography, object shapes are readily well represented. However, shapes, and their dimensions, provided by the synthetic aperture radar oblique view are often unlike and disproportional to the actual objects they represent. In addition to perspective distortions, various parts of an object may backscatter the energy in different ways. Objects and terrains that have an important vertical dimension may appear reversed due to image layover, or compressed by foreshortening, and, as earlier discussed, sometimes sided by radar shadows. Image layover occurs if the radar beam depression angle is larger than the exposed slope angle. This angular setting has the radar beam reach the upper portion of an abrupt hillside or building wall before it touches the surface below. As backscattering is imaged following this very sequence, a switched image of the high and low slope locations is created. This displacement is noticeable when the objects' image appears stretched and points to the near range direction. Foreshortening and specular-point features trigger a very strong backscattering from surfaces where the local incidence angle is, or near, zero. This effect causes the imaged object to appear compressed compared to the actual dimension, but this is difficult to visualize. Foreshortening, recognizable by a very bright tone, is produced by

FIGURE 3.13

Star-shaped bright tone features represent a very high backscattering received by the synthetic aperture radar system. Saturation causes surrounding column and row pixels to be affected, but in gradually lower magnitude away from the center. The condition for this to occur is that a particularly high reflective surface points or faces directly at the radar receiver (Raney 1998). Star shapes are agglomerated at the location of a vehicle storage lot (5-right,3). The proximity of metallic angular surfaces causes a very strong backscattering. It is frequent to have star-shaped backscatter from very small objects such as power line equipment or communication towers. For example, in Figure 3.12, one object is located about 50 m east from the lake (5-left,3). Image: CSK 908 HH 31 deg RA Cortona W 2.4 km LL 43.304957 11.930500 deg.

a direct return to the receiving antenna of the backscattered beam. Effects of image layover and foreshortening, with occasional radar shadow, impacts the understanding of vast, very abruptly sloped topographical units and high rise constructions. Smaller vertical objects such as fences, valleys, and trees are strategically enhanced by image foreshortening and shadow. These preferentially enhance linear shapes aligned perpendicular to the radar beam look direction.

3.6 Summary

This chapter has presented interpretation keys, traditionally considered for information extraction from aerial photographs, purposed for use with synthetic aperture radar image interpretation. Clues and evidence leading to land use and land cover descriptions were developed from systematically observing the image, texture, pattern, shadow, shape, and dimensions. However, these keys applied with optical data must have their connotations revised when used with images acquired from synthetic aperture radar systems exploiting the active remote sensing model, oblique perspective and long electromagnetic wavelengths. This chapter establishes the methodology for interpreting the LULC from radar images. Explanations of how the interpretation keys connect to technological concepts carry into examples that help the readers to familiarize themselves with the descriptions. A standard terminology is suggested to designate visually assessed characteristics from each key. Image specifications, a scale, and location are given at figure caption end.

4

Regional Land Cover Descriptions

Earth observation satellite images displayed with a small cartographic scale or simulating a high eye altitude view routinely offer a first glance observation of regions. The purpose of this chapter is twofold. First, it presents land cover and landscapes that extend over very large areas, of several square kilometers. The synthetic aperture radar image product complete path width and nearly all of the azimuthal extent frame the representation for each region. Second, it introduces the different regions among which large scale geographical ensembles were chosen for the next chapter on large scale geographical ensembles. Regions are presented through a geography narrative, an image, and an account of the interpretation key characteristics that help deduce land cover information. The section titles are meant to summarize the geographical context. Some of the mostly populated centers are then listed, with some background information. An overview of the land cover type for the area represented on the image will familiarize the reader with the levels of information that can be visually obtained from a document that is normally approached as spatially generalizing. The image representations are about equivalent to those provided from a ©2018 Google eye altitude of about 45 km. The image width, stated in the image caption, gives a sense of the detail that can be obtained before recourse to a large scale depiction.

CSK and RS2 images were acquired for seven regions of Canada and Italy. The Canadian locations include Fredericton, Happy Valley-Goose Bay in Labrador, Montréal, and Ottawa. The three sites in Italy are Cortona, Taranto, and Ferrara. Commonalities framed by the interpretation keys and background information about the locations facilitated identifying six of the Anderson Level I (Homer et al. 2007) land cover categories. 'Developed' is the most frequently represented type of land cover throughout the image set, followed by 'Forest', 'Agriculture', and 'Water'. Small areas of 'Wetland' and 'Barren Lands' are represented on a few images (Table 4.1). The visual examples show the land cover types in different geomorphological settings, spatial distributions, and seasons.

4.1 Coastal Development

Taranto is a large industrial center of Southern Italy (Figure 4.1). With a population of 200,000, it is the second largest city in the Apulia region. This port

TABLE 4.1
Land cover of predominant interest for the different regions.

Land cover	Location
Agriculture	Cortona, Ferrara, and Taranto
Barren	Labrador
Developed	Montréal, Ottawa, and Taranto
Forest	Fredericton, Labrador, and Ottawa
Water	Labrador and Taranto
Wetland	Labrador and Ottawa

city, opening on the Ionian Sea, has the greatest proportion of its waterfront industrialized. The most important European steelworks and a national navy base are the main occupants. Other activities involve industrial machinery fabrication, cement work, oil refinery and service industries (King 2015). The city of Taranto, is located on a crescent-shaped coastal plain that extends to the Southwest, from which the altitude gradually increases northward to 500 m highlands. Meandering fluvial systems and artificial channels are developed through lower terrace and a trellis pattern upland (Gioia, Bavusi, Di Leo, Giammatteo, and Schiattarella 2016). Dark topsoil and alkaline soils on Quaternary deposits support the Mediterranean coastal steppe and evergreen forest, respectively. A band up to 2 km wide of dense forest occupies the highlands between Taranto and Martina Franca. Forested patches alternate and concede to low-intensity developed at mid-elevations. Farmed fields exhibit relatively small, but variable sizes of 1 to 15 ha. Larger lots, to about 30 ha, characterize reclaimed coastal plains portions West of Taranto. Specialized wheat cultivation is the choice on higher altitude terrain of the Apulia region, while the lower elevation on the Coastal plain and Mediterranean climate is favorable to vineyard, orchard and olive dominated agriculture (Walker 1967). Modest size population centers are dispersed across the terrace and foothills, e.g., Castellaneta, population 17,100, Mottola, 15,900, Palagiano, 16,100, and Statte, 13,900. A small portion of the littoral is occupied by residential and commercial functions, including Lido Azzurro, a low-intensity touristic development.

4.2   Forest Wrapped Developments

Fredericton is the capital city of New Brunswick, an Eastern Province of Canada. The population of Fredericton is 58,200 and is concentrated along about 20 km of the Saint John River shorelines. Pristine and harvested forests, which cover eighty-five percent of the Province's territory, surround the city

FIGURE 4.1

Three units are represented in the area. The Ionian Sea to the south displays
a very dark tone and fine image texture, indicating a calm water surface (5,2).
The arc-shaped band contiguous to the sea is a marine terrace (6-7,7) gradually transiting to the Apennine southern-end foothills (5,10). Except for a
narrow band of open forest on the shoreline (4,5-high), the terrace is overall a
brighter tone and coarser texture than the highlands. The city of Taranto, a
high-intensity development (10,3), is highlighted by bright tones and intermediate textures. Round shapes and bright tones help locate communities [(4-left,
10-low), (6-left,10-low), (6-left,8-low), & (9-right,7-low)] scattered throughout
the agriculture-intensive landscape. Very bright tones enclosed in rectangular areas portray permanent crops with stakes in trellis, solar panel fields,
and covered crops (3,8). Side-by-side with radar shadow, bright tone sinuous
lines mark the river valleys radiating from the terrace to highlands [(1,9-10),
(5,9-10), & (8,10)]. A coarse texture and patchwork pattern for areas of
small and large agricultural fields, respectively, and overall a speckled pattern highlight intermittent developments throughout the heavily cultivated
terrace area. Herbaceous planted lands alternate with permanent crops, such
as vineyards and olive groves, and herbaceous natural vegetation. Forested
areas display intermediate to dark tones and fine texture [(4,6-low) & (6-
left,6)]. Image: CSK 055 HH 29 deg RA Taranto W 40 km LL 40.505415
17.036687 deg.

and very narrow bands of agricultural lands near the Saint John River (Figure 4.2). Its tributaries from the Northwest and Northeast, the Keswick and Nashwaak Rivers, respectively, are bordered by wetlands, agricultural lands and a scattered development. Fredericton is in a portion of the Appalachian physiographic region that spreads East of North America from the state of Alabama, United States, to Labrador, Canada. Small wetlands occupy depressions all through the hilly landscape. The bedrock is mainly sandstone, scored by a dendritic drainage pattern. At Fredericton the Saint John River has a width of about 600 m. The terrain up to 10 m ASL is a flood plain onto which a portion of the central business district (Broster 1998, City of Fredericton 2008) is located. Islands in the Saint John River spread over a distance of about 10 km between Mactaquac and Fredericton; they are uninhabited, agricultural or grass-covered fields. To the West, the River is intersected by the Mactaquac hydroelectric dam, and farther upstream the head pond rests at the historical river bed location (Stantec Consulting Ltd 2016).

4.3   Heterogeneous Land Cover

Ottawa, Ontario, is the Capital of Canada and neighbor to the city of Gatineau, Québec. Set apart by the Ottawa River, the urban areas extend over a radius of about 15 to 20 km. Ottawa, population 880,000, is on the South shore of the River, while Gatineau has a population of 270,000 (Figure 4.3). The cities' central business districts, government hubs and commercial functions are situated near the Ottawa River. Surrounding land cover types include residential, agriculture and forest. North, dense metamorphic and intrusive rock types shape the Boreal Shield Ecozone's Gatineau Hills, with altitudes from 60 to 400 m ASL. Jagged upper terrains and thin sediment-covered hillsides at lower elevations are grounds for black spruce dominated conifer forest, a few broad leaf stands, and sparse openings on exposed bedrock and wetlands (Urquizo, Bastedo, Brydges, and Shear (Eds.) 2000). Many ponds, lakes and rivers outline the rugged terrain topography. Small communities are scattered in the mountainous region; they are located on lakes and watercourse shores. A contrasting landscape just South is the softly undulating Ottawa River valley and tributaries, part of which is in the Woodland Plains Ecozone (ESTR Secretariat 2016). The best nutrient rich soils in this area are used for agriculture, while stream sides and less productive lands are occupied by a patchy forest cover composed of mixed broad leaf and conifer stands. Large wetlands are located West and East, relatively close to the Ottawa built environment.

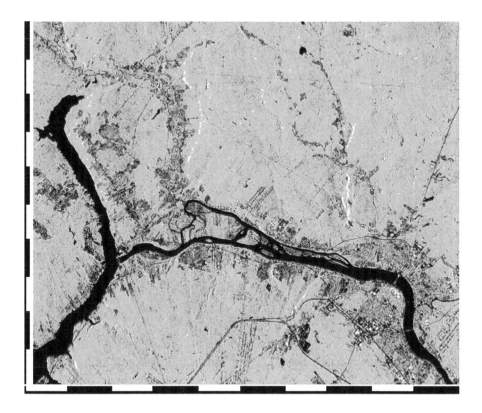

FIGURE 4.2
The developed land of the Fredericton city agglomeration is indicated by very dark, dark, and very bright tones forming a coarse texture and faint grid pattern, restrained along the Saint John River (9,2-4). The highest population density is indicated by a core of more contrasted tones and the convergence of roads [(9,3) & (9,4)]. Narrow bands of land developed for agriculture are confined to the Keswick River valley (4,6-9) and a few roadsides [(3,6) & (8,7-8)]. The Saint John River and the few small ponds scattered in the hilly terrain (7,10) are represented by a very dark tone and fine texture. Large forest expanses constitute about two-thirds of the land cover; they display a fine to intermediate texture and intermediate tone (4,3). Sporadic occurrence of curved bright lines marks escarpments, which are about perpendicular to the Saint John River [(4-right,6-9), (7,3), & (8,6)]. Elevation changes here are 30 to 100 m. Faintly contrasted straight lines across the landscape are electrical power transmission line corridors (9,1-10). Image: RS2 086 Four-polarization PCA 35 deg RA Fredericton W 30 km LL 45.995438 -66.779508 deg.

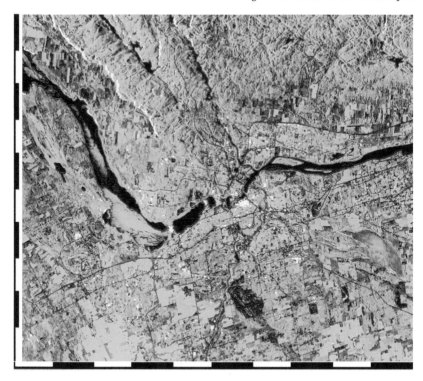

FIGURE 4.3

Contrasted layouts identify the Boreal Shield and Mixedwood Plains Eco-zones in the Ottawa-Gatineau area. North, the Shield is overall a pattern of crisscrossed bright and dark lines overlaid on intermediate tone and texture background characteristic of a dense forest cover (4-10,9-10). Entrenched valleys that are perpendicularly near the incident radar beam and one of the most important escarpments, marking a terrain rise of about 200 m at the Plains edge, cause jagged linear shapes and very bright tones. Narrow, sharply shaped and very dark tone rivers and ponds enhance the glacier-eroded bedrock depressions. Conversely, on Mixedwood Plains, flatter terrains are located on the wide and smoothly curved Ottawa (3-left,6), Rideau (6,4), and Gatineau (6,7) Rivers. Marsh and herbaceous wetlands, by their large dimensions and irregular curved shape [(9-10,4) & (2,6-7)], contrast with patchwork patterned croplands [(4,7) & (9,2)]. Medium- to high-intensity developed areas are in homogeneous sections throughout the Plains and near Ottawa River (6,5), some with a coarse texture, others a speckled pattern (4,5-high) and omnipresent dark tone road grid [(2-right,4) and (7,3)]. Very bright tone in patches identifies areas of turbulence on the Ottawa River waters (3-right,5), the commercial and business districts of Ottawa and Gatineau (6,5), and marsh vegetation (1,7). Dark tones in irregular-angular shaped areas identify as large open land lawns around airfields [(6-right,2) & (8-left,6)], and golf courses [(5,5-high) & (10-right,2)]. These are differentiated from dark tone lakes of varied shape [(4-left,8) & (8-left,10)]. Image: CSK 032 HH 18 deg RA Ottawa W 50 km LL 45.413093 -75.751879 deg.

4.4 Urbanized Islands

Montréal Island is bound by the Rivière des Prairies and Saint Lawrence River, in the province of Québec. Montréal, population 1.9 million, occupies the largest proportion of the Island (Figure 4.4). The land cover principally consists of very high-intensity development aimed at residential and industries. Other functions include business, public and institutional services, conservation parks, forests, and mixed use (Ville de Montréal 2019). Prominent landmarks include the Port de Montréal, which serves as terminal to the Saint Lawrence Seaway, the central business district, four universities, artistic and cultural institutions, and several green spaces. To the North is Ile Jésus that hosts Laval, with a population of 450,000. Across the Saint Lawrence, to the East is Longueuil, population 240,000. In addition to developments, Ile Jésus and Longueuil land cover includes agriculture, which is the Mixewood Plains Ecozone overall dominant land cover type (ESTR Secretariat 2016). The region's physiography is a layering of glaciolacustrine sediments at the ancient Champlain Sea base (Russell, Brooks, and Cummings 2011). Grounds are generally level and the soil is favorable to agriculture and mixed wood forest growth. In non-urbanized regions, corn crops and dairy mixed farming are the most common activities. In sparse forests of 60 % maximum tree crown closure, mainly grow sugar maple, poplar, aspen, birch, and spruce (Beaudoin, Bernier, Villemaire, Guindon, and Guo 2017). Terrain elevations represented in the image range between 20 and 60 m ASL, except the Mont Royal reaches 230 m ASL.

4.5 Valleys and Mountains

Cortona is located in the Tuscany Region, on the South-side foothill of the Septentrional Apennine Mountains (Figure 4.5). In this area, the terrain reaches 1,000 m ASL. The economy of Tuscany is based on agriculture and their derived producer industries (Walker 1967). Drained wetlands and alluvial deposits shape the reclaimed Valdichiana Valley grounds. Olive groves, permanent crops, heterogeneous agriculture, and arable land are spread over the gently sloped valley floor (Bernetti and Marinelli 2010). Large mosaicked fields spread the low lands while polyculture and small parcels occupy the hillsides near communities. Tall tree rows outline the boundary of several crop fields in the valley while forests cover the Apennine Mountains North of Cortona. As well, woodlands form ribbons or patches on the hillsides where it alternates with cultivation, rock outcrops, or winding roads. The population of Cortona is 22,500 and is somewhat unevenly distributed. Similarly to the other towns of Tuscany, very high-intensity developments are confined to the

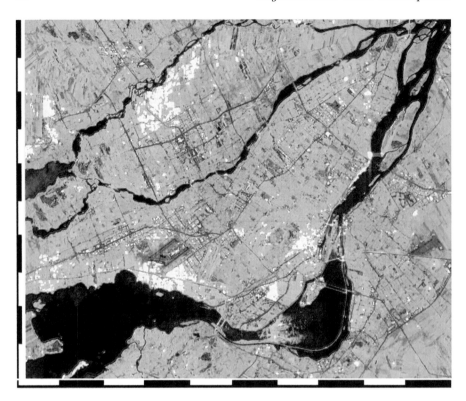

FIGURE 4.4
The Mixedwood Plains Ecozone level topography of the Montréal area facilitates the associated high-intensity development in a relatively homogeneous fashion. A sporadically contrasted grid of dark roads and streets creates an intermediate texture, overall representing residential land use. Service industries' buildings, and ancillary components are coarse texture and dotted pattern along wider arteries [(5,5), (5,7), & (9,2)]. Very bright tone areas distinguish residential neighborhoods where streets are perpendicularly aligned with the radar beam look direction [(4,3), (4,8), & (1-right,8)]. Wherever the bright tone area shapes are jagged or spread over any direction, they identify the tallest building concentrations (7,4), turbulent water of the Lachine Rapides (6-7,2), commercial centers [(4,8) & (9-left,8-high)], storage yards (1,7), and large industrial buildings (8-right,6). Very dark tones outline the river and canal's calm surface water, and help locate airports [(4,4) & (10,4)]. Dark tones in small irregular polygon shapes help identify lawn-covered urban parks and golf courses scattered across the region. Image: CSK 433 HH 22 deg RA Montréal W 45 km LL 45.540402 -73.676276 deg.

FIGURE 4.5

Land cover types in the Cortona area are in two particularly well contrasted units. A set of large v-shapes creates a pattern from a succession of radar foreshortening on hillsides exposed to the east and shadow tone darkening on west oriented slopes. The Apennine Mountains (6-8,6-8), as a group of sharp hills and valleys, is delimited from the moderate foothill (4,5) and even Valdichiana (2-4,2-4) topographies. Agriculture is the dominant land cover in these areas. Variable tones in coarse textures representing crop lands are mixed with a few forest lands (1-2,1-2), and a patchwork of large fields in the low elevation reclaimed lands (3,1-4). Water management channels are mainly oriented longitudinally to the valley's long axis, as well as are a railroad and several roads. These linear features differentiate from each other by a combination of their tone, width, and pathway. Channels are a little wider and display a bright tone bordered dark line (3,1). Roads are least contrasted with their intermediate tone but run throughout the landscape linking developed lands. Railroads are enhanced by a bright tone and a relatively straight path (2-right,7). Arezzo (3,9) is developed over a distance of about 20 km north of Cortona (6,3) and other population centers dispersed along the Apennine foothills. Relatively level, developed high lands are portrayed by a coarse texture and an interruption of the bright and dark alternate v-shapes [(10,7) & (10,9)]. Image: CSK 212 HH 31 deg RD Cortona W 40 km LL 43.352541 11.969281 deg.

community historical center, while scattered developments sprawl over arable lands at lower elevation. From 500 m ASL, the historical borough of Cortona overlooks the 5 km-wide Valdichiana Valley. The portion that is represented on the image extends from Northwest to Southeast of Cortona between Arezzo, population 99,500, and just North of Lake Trasimeno.

4.6   Widespread Agriculture

The land cover surrounding Portomaggiore, Italy, is predominantly crop farming on arable land (Figure 4.6). The communities of Ferrara, Portomaggiore, Molinella, and Lagosanto are located in the Southern Po River delta, which is a ribbon of thick alluvial deposits and surrounded by reclaimed lagoons of the VallidiComacchio Valley. This portion of the Emilia-Romagna Region is shaped by a gently inclined alluvial plain (Walker 1967); in the image area, the terrain elevation is just below 0 m East near Portomaggiore and increases to 7 m ASL on the West side. An intricate network of drainage channels and contrived rivers expands throughout the whole region. The farming industry developed harvest types that include sugar beet, hemp, cereals, and rice. These specialized crops, which occupy fields of up to 50 ha in size, are established on large expanses of reclaimed lands. The older properties at slightly higher elevations, which were not submerged before the time when lowlands were reclaimed, produce wine and fruit (Walker 1967) on small size parcels, seldom exceeding 10 ha. Throughout the region, isolated houses and farm buildings attest to agriculture dedicated lands. The integration to towns of small and medium industries facilitated economic development of rural areas during the second half of the last century (King 2015). The towns of Molinella, population 15,700, Portomaggiore, 11,800, and Lagosanto, 4,900, are high density Emilia-Romagna communities typically restricted to areas where the land is at least about 4 m ASL.

4.7   Wintertime Environment

Happy Valley-Goose Bay is a Northern community on the shore of Lake Melville and its tributary the Churchill River, Newfoundland and Labrador, Canada. It is a high-intensity development, with a concentrated population of 8,100 and covers an area of about 15 km^2. Forty kilometers to the Northeast, on the Grand Lake South and North shores, near its connection with Melville Lake, are situated the communities of Sheshatshiu Innu First Nation, population 1,300, and North West River, 550. The land cover for the imaged

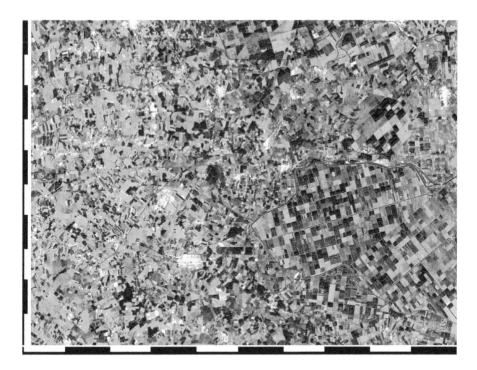

FIGURE 4.6
Population centers in the area of Portomaggiore (4-right,3) have a subtle display among this dense agricultural land cover. The patchwork pattern is made of large square fields in some areas (7,4), small crops in others (2,4), and variable contrasts throughout. From groups of small and irregularly shaped lots emerge areas [(4,5) & (4,9)] that locally correspond to higher terrain. Regular shaped fields are more common in the reclaimed lowlands. A water management channel system sets a rectangle fields patchwork pattern. Large regional channels enclose several lots [(8,4) & (6,10)] while smaller ditches may border individual crops. The larger infrastructure is represented by intermittent parallel bright pairs of lines, i.e., channel edges, with dark tone water in the middle. They expand regionally across the landscape and are sporadically bordered by town and cities. The smaller channels or surface waterlines are bright features bordering individual crops. Dispersed communities are represented by spreads of bright speckles along routes and channel shores. Spiked shapes created by buildings along roads radiating from community cores help locate small population centers, e.g., Tresigallo (5-right,8), Lagosanto (10,8), and Ostellato (7-left,6) with populations of 4,400, 4,900, and 6,100 respectively. Portomaggiore is a little better defined within a rectangular shape, partly outlined by a large canal. Image: CSK 006 HH 39 deg LD Ferrara W 40 km LL 44.746823 11.878899 deg.

area is forest-dominated, but interrupted by lands slowly recovering from a
1985-forest fire. A thick blanket of till that form drumlins or colluvium covers
the Boreal Shield (Ecological Stratification Working Group 1996) bedrock of
the uplands North of Melville Lake (Liverman and Sheppard 2000). The forest is principally composed of spruces, with a few insertions of low proportion
coverage by birch trees, poplar, and aspen (Beaudoin, Bernier, Villemaire,
Guindon, and Guo 2017). Happy Valley-Goose Bay is located on a sandy
plateau. Unconsolidated barren lands of the Churchill braided riverbed take
the form of alluvial island-shaped deposits and terraces (Liverman 1997, Liverman and Sheppard 2000). This wide curvy linear river comprises several
vegetation-covered islands. In addition to the large rivers and lakes, a multitude of jagged-contour ponds mark the bedrock while others established on
the Lake Melville lowlands are softly curved. Wetlands are scattered in the
lowland; a large one is just South of Sheshatshiu and others, smaller, along
the Melville Lake coastal plain and the Goose meandering river shores.

Winter time synthetic aperture radar images provide a unique potential to realize land cover assessments in the Happy Valley-Goose Bay area
(Figure 4.7).

4.8   Summary

The chapter Regional Land Cover Descriptions is an opportunity to present
the locations for which the synthetic aperture radar images were acquired
and to apply a LULC interpretation process to full-coverage images. Views
provided by thirty- to fifty-kilometer image widths serve land cover interpretation and some of the land use descriptions. Regional interpretation relies
on tones for representing specular, diffuse, and corner backscattering surfaces.
Tall buildings and uneven landscape configurations are particularly well contrasted with their surroundings. Bright tones consistently lead to land cover
with which components have vertical or slopes facing toward the incident radar
beam. Dark tones readily identify to specular backscattering on smooth surfaces such as calm water, barren lands, and dense broad-leaf forest canopies,
or radar shadow. Meanwhile, characteristic textures are difficult to consistently associate with particular land cover or land use types. The interpretation of image texture must consider that the spatial scale is generalized.
Several areas may display a fine image texture and others a pattern, with few
intermediate or coarse textures to exploit as interpretation criteria. Consequently, pattern, shape, and dimensions take precedence for the differentiation
between similar-tone LULC types. Grid and mesh patterns, regular geometric shapes, and small to moderate shape units mainly characterize developed
and human-managed land use types. Conversely, variable dimensions, irregular and rounded polygons, large recurring line-, v- or arc-shaped patterns

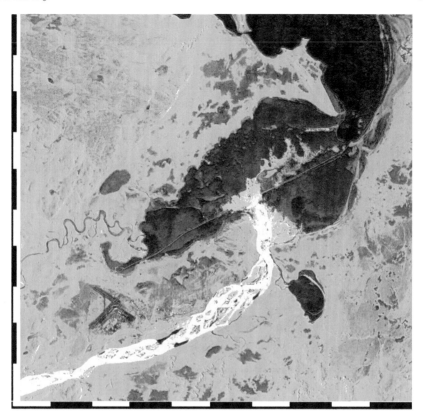

FIGURE 4.7

Acquired in early December, the image shows ice forming that had been encouraged by a month of below-zero Celsius temperatures. With a total depth of snow on the ground of 14 cm, major landscape components are clearly differentiated by their tone, from very dark to very bright, including water (9,10), ice-covered water (8,8), wetlands [(6,9) & (4,1)], barrens (3,7), dense forests (1,7), and ice floes (7,5). The texture is overall fine, apart from coarsely textured ice dispersing in Lake Melville (8,7) and intermediate texture with a notched pattern of the Boreal Shield surface. In this area, bright northwest to southwest oriented lines enhance abrupt slopes exposed to the radar beam and deeper bedrock breaks (1-4,8-10). Bright tone short ridges in lakes and the ocean give information about the forming ice scape. Developed lands comprise a large airport (3,3) and, displaying a speckled pattern, the town of Happy Valley-Goose Bay (4-5,3), and the communities of Sheshatshiu and North West River situated on the south and north shores, respectively, of the North West River where it flows into Lake Melville (6-right,10). These are connected by a road along one terrace edge discernible by association with bridges. While the ice is present, images give information about the navigation trajectories. Bright lines indicating broken ice in the ship wake lead to a wharf infrastructure (3,4). Image: CSK 208 HH 39 deg RD Labrador W 40 km LL 53.383695 -60.238629 deg.

of shadows and bright tones, and spatially integrated and consistent entities assist identifying landforms, topography, and naturally established land cover types.

The areas of interest in Italy and Canada are instrumental to demonstrating features of a coastal development, forest wrapped developments, heterogeneous land cover, urbanized islands, valleys and mountains, widespread agriculture, and in a wintertime environment. Each image features a dominant land cover type, and peripheral environments, including diverse developments, vegetation, water, wetland and barren lands.

The overall view, in different geographical contexts and proportional representations of land cover types, helps studying the visually rich information. In particular, the examples encourage adapting to various topographical settings, seasons, ensemble networks, and development intensities. Interpreted land cover in image sections and for particular items are located using a coordinate system set by the image graded frame. The figure captions expand on relevant interpretation keys, give coordinates for specific examples and list ancillary information including polarization, incidence angle, look direction and orbital path, place name, image width, and geographic coordinates.

5

Large Scale Geographical Ensembles

This chapter proposes visual examples of a relatively small expanse of land cover, land use, landforms, and object assemblages. Remote sensing applications' classification schemes and nomenclature used for the Atlas of Canada 2010 (Atlas of Canada 2010), the Coordination of information on the environment CORINE program (Bossard, Feranec, and Otahel 2000) and the United States Geological Survey USGS (Homer, Huang, Yang, Wylie, and Coan 2004) were the inspiration for the names given to the different geographical information items, by which the sections are titled. Nine main categories are considered, including Barren lands, Developed, Producer Industry, Service Industry, Landforms, Recreational, Vegetation, Water, and Wetland.

Each rubric includes a brief overview of what the geographic ensemble is about, followed by an explanation of the interpretation keys particularly relevant to identify the LULC from synthetic aperture radar images. Then an illustration is provided from one of the Cosmo-SkyMed and Radarsat-2 images that portrays a typical representation of the featured item. Figure captions identify examples of objects and LULC components and their (x,y) coordinate within the image according to bottom and left side graded axes. The ground width represented by the subsets is generally below 10 km. Background information helps put the interpretation in geographical and, for some cases, chronological context. The caption closing statements are image acquisition specifications useful to the LULC interpretation.

5.1 Barren Lands

Barren lands have their majority surface area composed of consolidated rock or granulated material. Compared to other types of land cover, flat rock outcrops with smooth surface textures are likely to present as dark tone and fine image texture. Due to the variety of ways barrens may be shaped and combined with different land cover types, the tone ranges from dark to intermediate. A compacted very fine grain size material or a veneer bedrock enables specular backscattering. Terrain topography related foreshortening and shadows may combine with, or override, the effects of barren material characteristics. Alternating with water, grass, or shrubs, the tone and image texture become brighter and coarse. The following sections present examples of a rock surface in urban environment and sand bars.

5.1.1   Rock

The image was acquired in 2014, at the end of a remediation project for the North section of the LeBreton Flats, Ottawa. At that time, the surficial soil had been removed for decontamination, exposing a large limestone-dominated bedrock face (Richard 1982), ponding, and very little vegetation (Figure 5.1).

5.1.2   Unconsolidated Material

Loose granulated sediments form a deposition environment, often along rivers, lakes, and ocean shores. Naturally occurring unconsolidated material may be buried or exposed, and spread over very large expanses of land such as deserts, dunes, and relic terraces. Exposed, with no overgrowing vegetation, and low moisture, the radar backscattering interacts with the surface according to the grain size of materials such as silt, sand, and gravel. Based on the modified Rayleigh Criterion (Peake and Oliver 1971, Sabins 1987), synthetic aperture radar X-band (Table 2.1) and C-band (Table 2.2) wavelengths are suited for differentiating sand from gravel. Material with grain size of less than 0.4 mm,

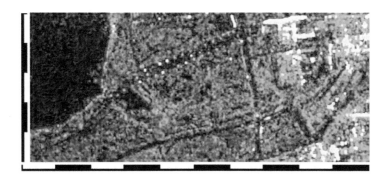

FIGURE 5.1
The medium to dark tone area rectangular shaped field represents the LeBreton Flats barren lands (5-6,2-3). Mottled texture is located where the uneven, but low topography, rock surface is exposed (6,2). The barren land area displays a moderately diffuse backscattering, which produces tones brighter than those of the wide asphalted roads (7,1-5) and Nepean Bay calm water [(1,3-4) & (4,3-low)], but very similar to grass-covered fields [(5-6,5) & (6,1)]. Ponding in the area shows in four very dark tone spots (5,3). Bright dot rows on LeBreton Flats' north edge denote city light posts alongside the Sir John A. MacDonald Parkway [(4,3-high) to (6,4-high)]. Image: CSK 032 HH 18 deg RA Ottawa W 1.1 km LL 45.414513 -75.715728 deg.

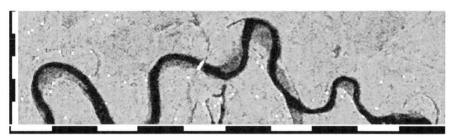

FIGURE 5.2
On the image, unconsolidated material displays a dark tone and intermediate image texture. The crescent shape [(2-left,3), (6-right,4), & (8-right,2-high)] and proximity to the river's very dark tone, fine texture, and sinuous shape indicates that the accretion zones are active, which prevents vegetation from establishing. The best tone and texture contrasts are obtained if the water surface is calm; otherwise, multi-polarization or multitemporal imagery may be needed to delineate unconsolidated material from rough waters. Despite a similar image texture, the accretion zones are well differentiated by their darker tone from the immediately surrounding forests [(1,4-5) & (6,1-2)]. Dark tone and fine image texture similar to the river crescents represent roads (5-left,1-5) and forest-cleared fields [(4,5) & (8,4)]; however, these are remote from the river and rectangular shaped. Image: RS2 474 FQ17 HH 37 deg RD Labrador W 7.2 km LL 53.393541 -60.415213 deg.

which encompasses medium and fine sand, silt, and clay (Blott and Pye 2012), will likely cause specular backscattering where these occur; a dark tone and very fine image texture should be expected. Coarser than 5 mm sediments, i.e., fine-to-coarse gravels and boulders, result in diffuse backscattering dominating. Unconsolidated material are part of meandering river beds in the form of accretion zones. These of the Goose River system (Figure 5.2) are lateral crescents and point bars on the meander bent inner sections. They are made of sand (Liverman 1997) and likely fine gravel (Robert 2014).

5.2 Developed

The presence of infrastructure built by people constitutes developed land cover. Assortments of construction shapes and materials and spatial distribution bring a heterogeneity of responses on a radar image. Common traits include frequent occurrence of very bright and bright tones for the buildings interspersed with intermediate to dark tones for the vegetation and streets, respectively. All these elements are arranged into grid or blocky patterns, and

land use-related ensembles often emerge due to the proximity of similar buildings or the element complementarity. Built landscapes are shaped by factors
such as the historical, cultural, economical and political contexts, the physical
landscape, and contemporary planning policies. Land cover as a variable is
generally categorized about the relative or absolute concentration on ground
surface of infrastructure as the expression of human presence on the landscape. Meanwhile, land use identifies the type of human activity for which
the infrastructure was built. Both the land cover and land use are considered throughout the book. This section addresses four land cover categories
referring to development intensity levels (Homer, Huang, Yang, Wylie, and
Coan 2004), and one of them, high-intensity, is expanded upon with land use
category examples for commercial, industrial, and residential.

5.2.1   High-Intensity

High-intensity developed land cover denotes nearly continuous built-up and
impervious surfaces, on 80 to 100 % of the ground (Homer, Huang, Yang,
Wylie, and Coan 2004). The image tones are from very dark to very bright,
the texture is coarse, and patterns are expected to be speckled, blocky or
gridded. Contrasted backscattering by buildings, objects and the surface usually follow a regular sequence. High-intensity developments often display a
large number of very bright tone point-like elements due to double backscattering onto angular buildings' faces made of very fine surface texture such as
concrete, metal, and glass. The city of Ferrara (Figure 5.3) founded over a
thousand years ago is located in the Northeast of Italy, on the Po River South
shore, about 50 km from the Adriatic Sea.

5.2.1.1   Commercial and Business District

Commercial and business districts display coarse texture and blocky pattern.
A collection of very dark to very bright tones is produced by mixed building
heights, shapes, and construction materials. Bright tones characterize corner
reflection and foreshortening of the radar beam on smooth glass and metallic
material of very tall buildings. Radar beam overlay effects on vertical walls
give an appearance of the high-rise buildings being geographically shifted. The
concentration of very bright tones and overextended rectangular shapes define
the densest city center. Contrasting dark tones occur more often where buildings are lower and allow specular backscattering on asphalt surface parking
lots and streets. Towers of the Montréal commercial and business district are
represented by a very bright tone, at the location of tallest buildings, between
the forest covered Mont Royal and the Saint Lawrence River (Figure 5.4).

5.2.1.2   Industrial

High-intensity industrial developments exhibit agglomerated very bright tones
on intermediate to dark background, alternating fine and coarse textures,

FIGURE 5.3

Row houses and streets alternate to form a grid pattern over the majority of the city of Ferrara. The street angles vary, in blocks, but the even building heights create a relatively homogeneous pattern as a whole. The high-intensity development is bordered by a railroad and triage yard (2,4-5), linear park (10,1-5), and canal [(1,3) to (5,1)]. Slightly darker tones, of a grid pattern, delineate a neighborhood that has a sparse tree cover and wider streets (4,4). Some of the very dark and fine textures represent the canal water, rectangular shaped sport field grounds (3,4), and small pockets of radar shadow [(8,4) & (9,1)]. Lawns, parking lots, and roads display dark tones. Intermediate and bright tones represent most buildings, while the cardinal effect appears to cause the large areas of very bright tones. In these, the streets and buildings are aligned perpendicularly to the radar beam look direction [(5,5) & (8-9,5)]. Tall modern buildings (2,3) and a communication tower (5-left, 1-high) also cause very bright tones. Image: CSK 806 HH 27 deg RD Ferrara W 3.3 km LL 44.833496 11.616779 deg.

and irregular blocky patterns. Very bright tones appear due to a high concentration of corner backscattering surfaces on metallic building walls and industrial equipment. The overall layout differs from commercial and business districts by the spatial heterogeneity of building shapes and height. Industrial sites often enclose large outdoor equipment, pipelines, vehicles, service roads, vacated grounds and storage yards. The industry type is a determinant of the land cover in terms of exposed barren grounds, vegetation, and buildings. Bright tones characterize smooth glass and metallic wall covering material, while dark tones represent radar shadows or specular backscattering,

FIGURE 5.4
Streets form a regular grid pattern that emerges well where the buildings
are shorter. Side-by-side radar foreshortening and shadow highlight the east-
and west-facing linear shapes, respectively, and indicate the forest-bordered
road sinuosity [(1,1) to (2,5)]. Corner backscattering and bright star shapes
are concentrated at the commercial and business district's tallest buildings
(6-7,2-3), industrial harbor front (10,1-2), and a monument (3-left,5). Other
objects such as a cliff (3,3) and residential buildings (4,4) that face the incident
radar beam produce linearly shaped intermediate to bright tones. Image: RS2
476 HH 35 deg RD Montréal W 4.5 km LL 45.502388 -73.572207 deg.

for example on roof tops made of asphalt or metal. Building shapes are well
outlined by foreshortening and shadows, particularly for large warehouses,
workshops and round or dome shaped storage tanks. Industrial activities are
typically found near a transportation axis, such as waterway, railroad or highway, and apart from or at the peripheries of communities. Immediately West of
Taranto, a petroleum and steel industry dominates the landscape (Figure 5.5).

5.2.1.3   Residential

A series of nearly same dimension dwellings, at even distances from each other,
form areas that are noticeable on synthetic aperture radar images due to contrasted high frequency tone changes, and meshed or gridded pattern. Neighborhoods display patterns that represent a repetitive and mirrored sequence
of backyards, buildings, front yards, and streets, often with trees and vehicles. Sections are usually perceptible within the residential fabric that from
one block to the next may change by the building dimensions and spacing.
In Montréal, a neighborhood is represented by bright and medium tone grid
patterns (Figure 5.6).

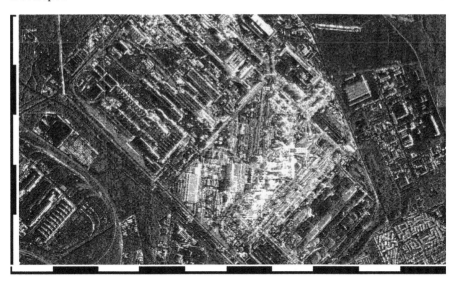

FIGURE 5.5
The plant includes a production core (6,2-3) of chimneys, pipelines, buildings, and reservoirs, and a sector (3-4,4-5) with low-lying hangars, parking, and outdoor storage areas. Close by is another industry type which has different building shapes, density, and layout (1-2,1-3). Overall, the industrial sites include very few roads. Radar beam foreshortening, as well as the large surface areas for buildings and outdoor spaces, helps distinguish the industrial land use from other high-intensity development types. For example, moderate and unevenly-dimensioned buildings identify to a commercial area (9-10,3-4), and similarly-shaped small buildings characterize a residential neighborhood (10,1). Image: CSK 840 HH 31 deg RD Taranto W 4.5 km LL 40.504089 17.211539 deg.

5.2.2   Medium Intensity

Built-up and impervious surfaces cover 50 to 79 % of the land (Homer, Huang, Yang, Wylie, and Coan 2004) in order to represent the medium intensity developed category. A mix of land cover types, including vegetation, undisturbed land, managed natural milieus, water, built-up, asphalted, and landscaped areas, contributes to the high spatial variability of tones, therefore producing intermediate to coarse textures and different patterns. The building-to-lot area ratio is low compared to high-intensity developments. As the climate allows, mature neighborhoods often have tall trees that overhang the houses, sidewalks, and streets. The outcome is a radar beam that, in reaching the tree canopy, has a large volume scattering component. As a result, the street grid pattern is attenuated on the image and a coarse texture occurs (Figure 5.7).

FIGURE 5.6

The succession of dot-like intermediate-tone objects represents houses in lots, which are bordered by dark thick lines and by thin lines for the streets and backyards, respectively. In sections [(4,1-low), (9,4), & (6,10)], house lots display much brighter tones likely because of the perpendicular geometrical alignment with the radar beam look direction, i.e., referred to as cardinal effect. Dark tone and fine image texture areas discontinue the pattern at the location of schools (4-right,2-high), parks [(2,4) & (5,3-high)], and recreational facilities [(8-left,1) & (2-right,3)]. At the boundary, service industries occupy bigger buildings [(1-2,1) & (6,2)], and in the center, a shopping mall (3,3) and parking lot (2-right,3) are connected to local roads and major arteries. A linear park by a canal [(5,1) to (9,5)] and trees along a river (10,1-2) display dark to intermediate tones, and intermediate to coarse textures. Image: RS2 476 HH 35 deg RD Montréal W 4.0 km LL 45.429923 -73.620296 deg.

5.2.3   Low-Intensity

Built-up or impervious surfaces cover 20 to 49 % of the land for areas where low-intensity development occurs (Homer, Huang, Yang, Wylie, and Coan 2004). The buildings are distant from each other, while vegetation, water bodies, or other types of natural grounds that let surface water percolate cover at least one-half portion of the lands. Bright tone marks the buildings on a medium to bright tone background for the surrounding expanses of land. A coarse texture is formed where the constructions are clustered, but if they are spatially distributed in a random or dispersed fashion, then a dotted or speckled pattern is created. This contrasts with the relatively even texture of surrounding land cover types. Lot boundaries being often apparent result in a blocky pattern. Because low density developments sometimes unevenly cover very large areas, categorization into the low, compared to intermediate,

FIGURE 5.7

A residential neighborhood of the Ottawa Region has the characteristics of medium intensity developments' land cover due to the large lot on which the houses sit, and the surrounding ample shrubs, grass, and tree cover [(4,3) & (5,4)]. A faint grid and sharp speckle pattern are formed by dark and very bright tones. The high backscattering components, i.e., houses and accessory buildings, are at greater distances from each other than for the higher intensity residential area to the east (10,3). Few curved roads cut through intermediate tone and intermediate-to-coarse texture forest (2-3,1-2) outline a hilly terrain. Darker tones and finer image textures represent school yards (6,2), outdoor sport grounds (7-left,4), grass (9-10,4), a wide street along the river shoreline, and water [(1-2,4) & (9,2)]. Of very bright tone, larger features are schools [(6,1) & (5,2)], a warehouse (7-right,4), and a wharf (7,5). Image: CSK 909 HH 18 deg RA Ottawa W 2.0 km LL 45.453119 -75.674420 deg.

intensity is sensitive to the considered interpretation unit size. Buildings may serve as focus. By evaluating their relative size and that of the immediately surrounding grounds, it more easily leads to a percent cover estimate and land cover type identification. Linear developments, for example in coastal zone and along regional roads, may sporadically form a medium or high-intensity developed territory. With a smaller interpretation unit, an area may fall into the medium intensity category where the buildings are concentrated, while by moving to a regional scale, a low-intensity development may eventually fit the open space category. The West end of Castiglione del Lago, Italy, has a low density population, with houses built along a regional road and few other narrow transversal streets (Figure 5.8).

FIGURE 5.8
In an area where the land is given a mainly agricultural function, low-intensity developments prevail [(2,1) to (7,3-low), & (4-5,3-4)]. Gradually changing to medium intensity (7,3), the developments further increase in density toward the east with land for service industries (8,4-5) and residential developments (10,3-5). The regional road is noticeable as a dark tone straight line [(1,1) to (10,4)] that passes underneath (8,4) a double-railroad displayed as a wide bright tone line. Image: CSK 212 HH 31 deg RD Cortona W 2.1 km LL 43.123434 12.022528 deg.

5.2.4   Open Space

As part of the developed land cover, open space is a category that represents areas with the least proportion of human built infrastructure. This land cover type may be part of, or in proximity to, the urban environment but not including any significant construction. Buildings are scattered throughout managed and natural environments as impervious grounds occupy up to 19 % of the area (Homer, Huang, Yang, Wylie, and Coan 2004). The image tones are dark, intermediate or bright, depending on the surface characteristics. The image texture is generally fine to intermediate because very few buildings and infrastructure bring contrasted backscattering to the image. Yet land use of open spaces by humans may regulate the image pattern. Tone and texture differences, that are land ownership related, create geometric shapes and block or patchwork patterns. Otherwise, large expanses of natural undeveloped land may display homogeneous fine, intermediate or coarse textures, and patterns related to factors such as soil and vegetation types, or topography. A maple tree and red oak closed forest canopy dominates the open space provided by the Mount Royal Park, Montréal (Figure 5.9).

FIGURE 5.9
The image encompasses open spaces dedicated to cemeteries [(1-2,5) to (6,5)], forest lands [(6-7,1) to (8-10,5)], and a lake (5,1). A small area shows low-intensity development (9-10,1-2) on the mountain foothill. Deciduous tree dominated forests displays a coarse texture. Overall the tone is bright compared to that of grass-covered cemetery lots. Bright tones and coarse texture also characterize the trees in rows and patches throughout the grounds. A large parking lot shows a faint striped pattern (6,2) and the lake is represented by a very dark tone and fine image texture. Sparse buildings [(1,5) & (6,5)] and a communication tower (7,4) display as bright pointed shapes. Foreshortening effects brighten the west exposed mountain slope (7,4-5) and radar shadow dims the opposite side (10,4) given by a right look from ascending orbit beam. Image: CSK 433 HH 22 deg RA Montréal W 2.0 km LL 45.502508 -73.596634 deg.

5.3 Producer Industry

Producer industries involve, for example, growing crops, mining natural resources or processing raw materials from the Earth. These activities are often exposed in a manner that makes them noticeable using remote sensing. This section offers examples of synthetic aperture radar images of land use related to the industries of agriculture, hydro power, extractive, and resource transformation. The landscape is strongly modified by most of these processes but the land cover typically remains permeable and includes few buildings. Therefore, producer industries often match low-intensity developed land, or open spaces in terms of impervious surface proportional coverage.

5.3.1   Agriculture

Agriculture consists of land based arable and pastoral farming practices. The occurrence of different land cover elements is contingent on factors such as product type, built infrastructure that is required for supporting the activity, yield, i.e., subsistence or commercial, and whether produce is transformed and packed on site. A common trait is the large proportion of open space it occupies. This landscape may be identified as low, medium or high-intensity development where it requires buildings. Agricultural fields display a variety of tones and medium to fine image textures, and simple geometric shapes arranged into grid or patchwork patterns. Factors such as crop type, moisture state, and growth stage regulate the tone and texture. Regular size and shaped fields are often found on gentle terrain slopes and they display homogeneous tone and texture. Infrastructure that may cause a very bright image tone due to corner backscattering includes houses, barns, silos, and various specialized function buildings such as a greenhouse, chicken coops, carousels, and machinery. All of these would be concentrated at one or a few locations within a property. In addition, fences and water management channels and pipeline networking through many fields form line shaped features. Because they are vertical or dug structures, they change the terrain surface shape. Otherwise, terrains supporting agricultural fields are ideally level. Linear features, also including vegetation bordered roads and in-field trails, are often enhanced on the synthetic aperture radar image by bright and dark tones due to radar beam foreshortening and shadow, respectively. Crop fields West of Cortona (Figure 5.10) display bright to dark tones and an intermediate texture.

5.3.2   Hydro Power

Hydroelectric energy production relies on natural waterways, whether they are small streams or very large rivers. The water flow energy is harnessed and transformed into electrical power. While a significant portion of a hydro power plant is underground or immersed, nevertheless, this type of construction makes structural and environmental modifications to the landscape. A dam perpendicularly crossing a waterway, upstream reservoir, reshaped riverbed, power transmission line right-of-way and substations are all apparent on the Earth's surface.

5.3.2.1   Dam

A hydroelectric dam forms a barrier transversally to a river. A large dam may have a noticeable curvature, concave upstream, or a straight shape. Corner backscattering and foreshortening of the incident radar beam on concrete and metallic material causes a strong return, contrasted with the dark tone of surrounding water bodies. In some cases, exposed bedrock or turbulent waterfalls on the dam's downstream side display an intermediate to bright tones,

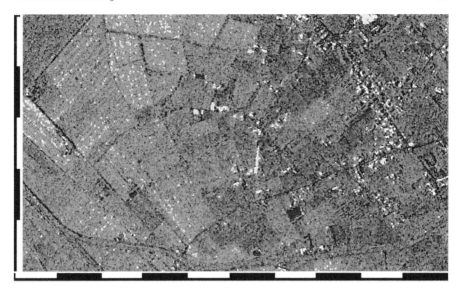

FIGURE 5.10

A striped pattern identifies fields with water control channels or plowed rows (2-3,4-5). Very dark tones of small rectangular and square shapes represent irrigation ponds [(3,2) & (7,2)]. The roads have a dark tone over linear shapes, some poorly contrasted, and they link the open spaces to a low-intensity development neighborhood (9,3-4). Herbaceous vegetation-covered drainage channels are distinct by their line shape corridor, bordered by thin bright and dark lines (1,2). Clustered farm buildings and houses (5,4-low) and a metal fence (6,3-low) are noticeable by their bright tone. A few forest patches [(7-8,1) & (8-9,4-low)] and field-bordering tree rows [(4,5) & (9,1)] are well outlined by very dark radar shadows, but the tree crowns themselves display dark to intermediate tones that mix with the crops. Image: CSK 908 HH 31 deg RA Cortona W 2.5 km LL 43.269743 11.916447 deg.

as a result of foreshortening and diffuse backscattering, respectively. The arc-shaped Mactaquac Dam (Figure 5.11) harnesses energy from the Saint John River 20 km upstream from Fredericton. It has a production capacity of 670 MW (Stantec Consulting Ltd 2016).

5.3.2.2 Reservoir or Head Pond

A very dark tone and fine texture depict a water-filled reservoir upstream from a hydroelectric dam. The artificial broadening of a river upstream from a dam makes the channel and tributary valley junctions appear wider and have spiky shapes in contrast with those of natural river segments. Rapides-Farmer,

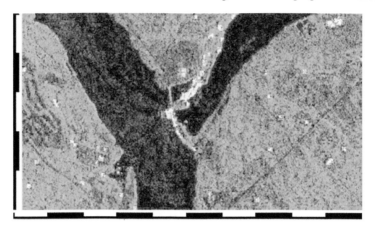

FIGURE 5.11
The dam infrastructure (5,2-3), which is also a bridge for Mactaquac Road, is outlined by bright tones at the location of a metallic rail guard and over-hanging light poles. Saint John River displays a very dark tone and fine image texture. Immediately by the dam, the tone is bright due to either exposed con-crete or rock face. On an image acquired at another time, a turbulent water outflow may cause diffuse backscattering at this location. The power gener-ating station (6,4-low) and power transmission substation (5-right,4) display bright and very bright tones. A rectangular and almost 200 m in length poly-gon represents the area occupied by the generator station, which also is located closely to the dam. The substation is smaller, but this all-metallic installation is greatly contrasted with the dark-tone asphalt and barren grounds. Image: RS2 087 HH 28 deg RD Fredericton W 4.1 km LL 45.952655 -66.868613 deg.

upstream, and Chelsea Dam hydroelectricity generation plants on Gatineau River (Figure 5.12) have 100 and 150 MW production capacity (Hydro Québec 2019), respectively.

5.3.2.3 Transmission Line Corridor

Transmission line corridors, or rights-of-way, exhibit straight linear features whose width is determined by factors such as the pylon size and voltage grade. This power transportation infrastructure runs across about any type of land-scape and often parallels roadways. It greatly differs by the tone, texture and shape from a forested land cover. A forest displays intermediate to bright tones on radar images that contrast with the transmission right-of-way dark tone. Tree clearing is a maintenance practice that reveals a low backscattering due to grass cover and shrubs. Through a forest, very bright and very dark tones enhance corridor edges where foreshortening and shadow are displayed in parallel, optimally with a right angle orientation of the incident radar beam.

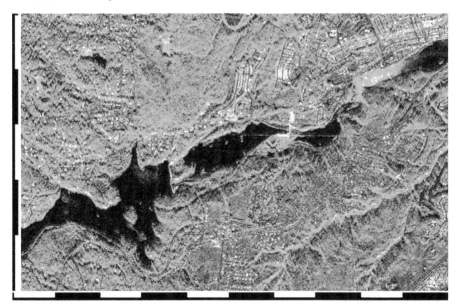

FIGURE 5.12
The reservoir widens river flow in valleys. The shape upstream [(1,1) to (4-left,2)] is characterized by sharp inlets on both shores, compared to the smoothly curved river banks downstream [(8,3) to (10,5)]. They display an even image texture and dark tone, for a calm water surface. Bright tones and a intermediate texture downstream (9,4) indicate shallow, turbulent water. High backscattering and sometimes star-shaping (7,3) are represented at the location of buildings, transmission substations, and dams. The latter are linearly shaped and perpendicular to the river stream [(4-right,2-3) & (7,3-high)]. Scattered bright dots in the forest landscape [(2,3) & (6,1)] are houses and other buildings that constitute low-intensity developments; one of them bordered by v-shape valleys [(7,1) to (9,2), & (9,1)]. The developed land's faint grid pattern of streets are in contrast with the intermediate texture forests [(8,1) & (4,4)]. Image: CSK 034 HH 18 deg RA Ottawa W 6.5 km LL 45.508010 -75.771368 deg. Note: Image rotated for display; north is to the left.

The metallic surface of pylons and some of the equipment they support tend to produce localized very bright tones (Figure 5.13).

5.3.3 Logging

Industrial timber cutting creates large dimension rectangular shapes on the landscape. Clearcut areas have a lower backscattering than other forested areas; medium tones indicate forest regrowth, and diffuse backscattering characterizes mature tree canopies. The tone of an eight-year-old regrowth is

FIGURE 5.13
Side-by-side wide and narrow transmission line corridors are joined by a third
path to form an upside down T (6,3-high). The portions through forestlands
are highlighted by radar beam foreshortening against the exposed tree and
shrub edges, and shadow on the opposite border, facing east. Other linear features in the image are a four-lane highway, narrow roads, and collector streets
of a low-intensity residential development (1-6,1). These asphalted surfaces
are of a characteristically very dark tone and fine texture. The overall homogeneous forest's intermediate texture is scored by a few bright tone lines; one
of them matches an opening on a trail (8-9,3). Image: CSK 936 HH 41 deg
RA Fredericton W 1.4 km LL 45.992015 -66.641638 deg.

expected to be similar to that of uncut forests (Ahern, Leckie, and Drieman 1993) possibly due to the canopy is covering sufficiently for debris to less
affect the incident radar beam (Murtha 2000). Several elements may affect the
backscattering and cause inconsistent tones in logging lots, including for example woody debris, organic material decomposition, soil coarseness, vegetation
regrowth, and wetness. This variability of tones translates into intermediate
and coarse image textures. Unless recently cut areas were well cleared, debris,
uneven grounds and tree stumps may coarsen the image texture. Contiguous
areas of different growth stages or management practices are spatially distributed in identifiable patterns such as checkerboard, patchwork, or stripes.
These emerge, for example, from selective logging practices that are meant to
facilitate resource reestablishment. A hostile topography hosts irregular field
sizes, shapes and patterns. The forest displays intermediate and bright tones.
Dark lines at regular distance from each other make a stripe pattern inside
many fields. Dark lines frame the forest patches that differ in height or that

FIGURE 5.14
The topography allows for managing relatively large lot sizes, of about 50 ha [(2-3,4-5), (3-4,1-2) & (7-8,1-2)]. Close to the river, on sloping terrain, the fields are smaller (6-8,4-5). In the image's western section, a power transmission line is well contrasted, with parallel bright and dark tone lines (1,3-5) as it runs across the forest stands. Similarly, logging roads to access the lots are much narrower and systematically located at the boundary of fields. In the valley, crops display a dark tone and intermediate texture, while the river displays very dark tones (10,5). Image: CSK 906 HH 24 deg RD Fredericton W 3.8 km LL 46.193286 -66.678579 deg.

have logging roads along sides. The logged grounds are distributed on undulating highlands South of the Tay River Valley (Figure 5.14), 30 km North of Fredericton.

5.3.4 Mining

Open extractive industrial sites display a wide range of tones and mostly a fine image texture. Tones represent the surface material grain size, which is usually very fine. Therefore, the tone is expected to be dark overall. However, quarries are multi-level, and as a result, radar shadow outlines terrace edges, specular backscattering occurs on flat sediment piles, ponds and roads, and bright tones highlight terraces, edges of sediment piles and fences facing the incident radar beam. Corner backscattering on buildings, mechanical equipment, and conveyors also cause bright features. The changing topography creates irregular patterns and shapes, except for a typical succession of concentric terraces from outward high level to inward low level of the excavation (Figure 5.15).

FIGURE 5.15

Here is the site of a sediment extraction and concrete production in Gatineau, with the excavation located in the northwesterly section of the property (3,2-low). A linear radar shadow outlines terraces along three sides [(2,2-4), (4-7,2), & (8,3)] of the excavation. Forest patches are also enhanced by faint shadows (9,2-4). Crescent shape radar foreshortening highlights grouped sorting piles (6,4). The lowest site of the excavation is indicated by rectangular and circular very dark tone polygons. These are water-filled collecting and intermittent ponds. The property is bounded by roads and a power transmission line; the latter is noticed by the bright speckles and wider right-of-way. Houses are scattered along one of the roads (4-10,1). Forests occupy the neighboring areas in all directions except south; it is of overall brighter tone and coarser texture than the quarry. A very high-intensity residential development is on the southern image edge (10,1-4). Image: CSK 909 HH 18 deg RA Ottawa W 1.9 km LL 45.433586 -75.845814 deg. Note: Image rotated for display; north is to the left.

5.3.5 Oil Refining

Oil refining industrial sites display a particularly large concentration of very bright tones, due to a multitude of corner reflectors. Distortion with blurred shapes result from a mix of uneven metallic buildings and objects. Dark tone areas near some of the very bright tone buildings represent radar shadow,

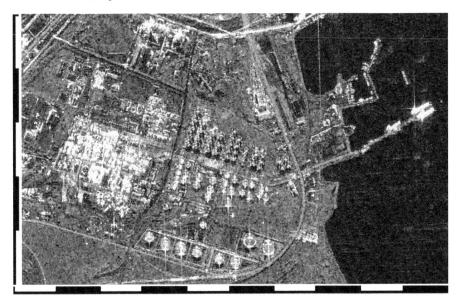

FIGURE 5.16
Compacted distillation towers, storage tanks, and pipelines create a bright tone and coarse texture (2-3,5). A speckled pattern represents a group of small reservoirs and rectangular buildings (3,3-4), while nearby large dark blocks are asphalted areas (2-3,4). Aligned reservoirs create a regular pattern of disc shapes (4-6,1-3), most of which are radar shadow outlined. Their concave and bumpy [(4-6,1) & (5-6,2-high)] or convex and smooth metallic (5-6,3) roofs are differentiated by the bright and dark tones, respectively. Fences, railroads, and roads border the industrial property and are further compartmentalized by retaining walls and a complex gas pipeline network. Three ships, noticeable by their very bright tone, are moored alongside one of the quays (9-10,3-4). Image: CSK 840 HH 31 deg RD Taranto W 3.1 km LL 40.485612 17.193971 deg. Note: Image rotated for display; north is to the left.

and medium-to-dark tones constitute the surrounding ground. Overall, a site has noticeable access to large water bodies, harbor facilities, and ground transportation, all of which are critical for the raw and refined product shipments. In Taranto, a large portion of the industrial harbor front is dedicated to the shipping and transformation of petroleum products (Figure 5.16).

Refined oil product storage tanks installed near communities provide energy supplies that must be periodically replenished. An oil tank park (Figure 5.17) is located about halfway between a harbor opening on the Atlantic Ocean and Happy Valley-Goose Bay. The ensemble is set apart from the community and uses a relatively large terrain with compartments set for several reservoirs.

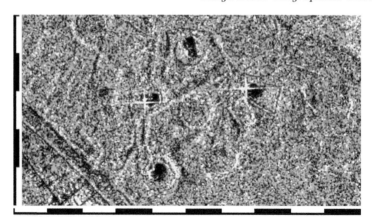

FIGURE 5.17
Four concave-top oil tanks display a very dark tone and near circular shape. The metallic construction facilitates corner backscattering [(4,1) & (5,5)], or star-shape bright [(4,3) & (7,3-4)] returns. Surrounding each tank is a retaining wall marked by a radar shadow. These safety basins may be circular (5,1) or of any geometrical shape (8,3). A couple of them were built [(6-7,2) & (5-6,3)], and, being unused, were overgrown by grass and shrubs. Image: CSK 925 HH 32 deg RA Labrador W 2.0 km LL 53.331486 -60.398658 deg.

5.3.6   Pulp and Paper

Pulp and paper production displays a high concentration of large low lying buildings on a property that is partitioned for the different industrial tasks. Areas are occupied with storage silos, residue piling, sedimentation tanks, or transportation infrastructure (Figure 5.18). The tones and textures are contrasted throughout and varied size geometrical shapes characterize such a site.

5.3.7   Water Treatment

Modern water treatment sites exhibit large circular basins for water sedimentation and aeration. Other processing and technological equipment are in buildings, whose dimensions are to some degree linked to the volume of water being processed. Water treatment facilities are more easily identified if they include open sky basins that display very dark tones due to the calm water surface. Each tank has a narrow overpass, a walking bridge, from the edge to center that produces a bright tone from the metallic construction material (Figure 5.19). Water treatment plants are sometimes integrated to the urban

FIGURE 5.18
A forest product transformation plant in Gatineau has newsprint as the main manufactured good. The property is occupied by a few very large buildings [(6,4) & (7,3-4)] that are linked by shafts [(5,3) & (6,2)] leading to outdoor sorting piles (4,3), small silos (6,2), and water treatment tanks [(6-right,1) & (8-left,1)]. Storage of heavy material, small buildings and power transmission substations are scattered on the property. They produce high and sharp backscattering compared to the intermediate to dark tone and fine image texture at the location of the grass-covered front yard. A forest (3,2-4) separates the plant from a group of heavy equipment service industries (2,2-5). It has a texture a little coarser than nearby grounds and radar shadow marks its boundary. A four-lane highway, high [(9,5) & (9,2)] and medium (9,3) intensity residential developments, and the Ottawa River to the south border the pulp and paper plant. Image: CSK 909 HH 18 deg RA Ottawa W 1.7 km LL 45.479895 -75.657107 deg.

area or dedicated to an industrial plant and in proximity of a water body, a river or lake to which the processed water is reverted.

5.3.8   Wind Turbines

A wind turbine is a stand alone energy producing metallic device that can reach a height of 150 to 260 m (Roberts 2019). The most visible components are rotating blades attached to a tower. Wind turbines are built often as part of wind farms, where they are aligned or form a grid, on land or off shore. Northwest of Cortona, wind energy is harnessed using equipment punctuating the agricultural lands (Figure 5.20).

FIGURE 5.19
The Robert O. Pickard Environmental Centre, Ottawa, comprises sixteen
open basins distributed in sets of four. Each basin has a standard diameter
of about 50 m. The intermediate and bright tones between the tanks and on
the west side of the property indicate the location of buildings [(4,1) & (3,4)],
closed tanks (8,3-high), and surface-hugging equipment [(3-4,2-3) & (6,3)].
Lawns and shrub landscaping [(2,5) & (5-6,1)] and tree rows [(6-8,5) & (2,1-2)]
create buffers at the property outskirt. Image: CSK 909 HH 18 deg RA Ottawa
W 1.0 km LL 45.462078 -75.588766 deg.

5.4   Service Industry

Service industries include distribution, food, retail, professional-scientific-
technical (PST) activities and transport (Government of Canada 2014). Storage and sales of tangible products destined for the public and delivery of education necessitate indoor facilities. There are similar requirements for PST
services, but in this case a relatively large number of employees must be
given workspace and, usually, parking areas. Service industries are established
within a variety of development intensities and land use types, in proximity
or easily accessible to the relevant customer base. Transportation services, as
a distinct category, are predominantly composed of outdoor infrastructure.
Networks are observed over large expanses of land and nodal constructions
create access for the potential customer base.

5.4.1   Indoor Service Industry

Indoor service industries occupy large, often clustered, buildings. This gener-
ates a blocky pattern of dark, intermediate, and bright tones, contingent on the

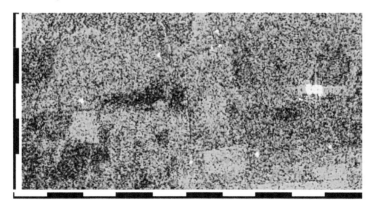

FIGURE 5.20
Wind turbines display a very bright tone restricted to very small point-shape features. The brighter tones for six of them correspond to larger towers and blades [(2,3) to (6,5), & (6,1) to (10,2)]. The six others are four times shorter, in groups of three at [(6-right, 4) & (6-7,2)]. All structures are aligned in the landscape on a southwest-northeast axis with a greater distance separating the large turbines from each other than is the case with the small ones. Bright objects, some star-shaped, include buildings and farm equipment (9,3). However, these are grouped and near roads. The background is herbaceous planted vegetation and permanent crops, typical of an agricultural industry land use. Image: CSK 055 HH 29 deg RA Cortona W 1.8 km LL 40.656137 16.887344 deg.

roof materials. Dark lines, due to radar shadow, and parking lots enhance the edge of simply square- and rectangle-shaped buildings. Conversely, they may have walls contrasted by bright tone, from radar beam corner backscattering and foreshortening. Communication towers and antennas, power transmission lines, pylons, and parked vehicles create very bright point features on the image. The variety of ways to make retail, food, education, and PST services functional and available result in different setups. They may constitute part of a traditional commercial business district of a large city, intensify residential developments, or shape urban sprawl. While these occupy specific spaces on the landscape, other service industries, such as virtual communication-supported trades, are out of sight. Images that follow give land use examples of mix sales (Figure 5.21), a PST service hub (Figure 5.22), retail (Figure 5.23), schools (Figure 5.24), and warehouses (Figure 5.25).

5.4.2 Transportation

The type of transportation system, for instance by air, water, or ground, defines the infrastructure elements and layout. Airports and harbors require

FIGURE 5.21
Wide roads in the west end of Arezzo, Italy, access a variety of food, leisure, retail, financial, and artistry services, and few manufacturing enterprises. The buildings are uneven in size and in very close proximity to each other. They are represented by the intermediate and bright tones and separated by streets or parking lots, which are dark toned. Bright lines on the radar beam facing the wall sides and a dark border for radar shadow outline the blocky pattern elements [(2-5,2-5) & (7-8,4)]. Comparatively, residential neighborhood buildings are represented by smaller and redundant shapes. Among them are a group of three- to five-story apartment buildings (6-7,2) and a block of two-story houses (9,3). Dark tone coarse texture along a curved axis indicates a treed river buffer (1-3,2) and a bright thick line, also curved, is a railroad and station (8-9,2). Image: CSK 212 HH 31 deg RD Cortona W 1.9 km LL 43.465693 11.851163 deg.

a large concentration of enclosed buildings and outside equipment. Airports occupy large expanses of land, while lake, river and ocean waterfronts host harbors. Structures for ground transportation, the roads and railroads, are linear features on the landscape. Roads display as dark lines due to specular backscattering on compacted surface material. Roads are distinct by their constant width, intersections, and integration to a developed land, i.e., the driveways and buildings to which they lead. Higher contrast with the surrounding landscape occurs while a road is parallel to the radar beam look direction because, this way, surrounding objects are less likely to obstruct the specular backscattering. Conversely, a road oriented perpendicular to the radar beam, if not directly exposed to the radar beam, may be identified due to roadside bright and dark tones, as a result of radar beam foreshortening and shadow, respectively. These can be caused by objects such as ditches, guardrails, trees or buildings flanking a road. The connection of different road sizes or direction involves interchanges and intersections. Crossings also require bridges and

FIGURE 5.22

Located west of the Ottawa city center, the Kanata North Technology Park is a hub to over 500 research and technological development companies in the areas of information and communication, manufacturing, health and life sciences (Kanata North Business Association 2018). Several square and rectangular buildings, up to about six stories high, are joined by narrow streets [(4-5,3-4) & (7,1-4)]. Parking lots alternate with the buildings, and landscaping includes lawns and forest patches. Dark tone wide lines are roads to access the site. One of the buildings, which displays a very bright tone and the appearance of layover, is a tall hotel (7,5-low) by a large parking lot (6,4) and a pond (7-right,5). A golf course (9-10,4-5), shrub and marsh open space (10-left,2-3), and a power transmission substation (4-5,1) are integrated but peripheral to the PST service industry area. Farther outward are high-intensity residential developments (1-2,5) and a deciduous uneven forest (1-2,1). Image: CSK 909 HH 18 deg RA Ottawa W 2.8 km LL 45.343236 -75.919270 deg.

overpasses. Railroads have a metallic and raised construction that contribute to backscattering of the radar incident beam. It displays a bright tone compared to other transportation axes or transmission lines. Railroads indifferently run through almost all types of landscapes, following wide radius curves or straight paths. They intersect roads and often reach the water front and industrial areas for transit of goods, and business centers for people commuting. The synthetic aperture radar image examples depict an airport

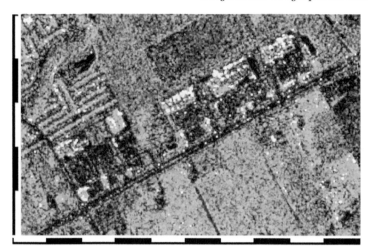

FIGURE 5.23
A group of contemporary big-box stores [(2,2) to (8,4)], and the nearby residential area, were built in place of agricultural parcels similar to those of the adjacent lands [(7-10,1-2) & (5-6,1)]. Flat ground facilitates the construction of various building sizes, in groups, one or two stories high, served by large parking lots, which are dark tone and lightly specked by either cars or lamp posts. The whole setting is spacious, contrasting with the very high-intensity dwelling nearby (2,3). The bright tones highlight the radar beam exposed building walls and the rooftop's protuberant ventilation outlets. Intermediate tones denote the box-store roofs and surrounding cultivated fields. Dark tones represent asphalted surfaces of parking areas and streets, tree line shadows, barren lands, and ponds. Image: CSK 909 HH 18 deg RA Ottawa W 1.4 km LL 45.287158 -75.908093 deg.

(Figure 5.26), port infrastructure (Figure 5.27), and ground transportation ensembles comprising bridges (Figure 5.28), interchange (Figure 5.29), road overpass (Figure 5.30) and roundabouts (Figure 5.31), and finally a railroad (Figure 5.32) and rail yard (Figure 5.33).

5.5   Landforms

Topographical features are detected, to a large degree, owing to the oblique view from which synthetic aperture radar images are acquired. The terrain slope influences the appearance of objects because it applies a local incidence angle, which sometimes greatly differs from the emitted beam angle of

FIGURE 5.24

Primary and high schools consist of rectangular buildings, either standing alone or as multi-modular units linked by aisles, whose dimensions depend on the student population. Building roofs display bright to intermediate tones on intermediate to dark background tones for lawns, parking lots, and sport grounds. Educational services are often part of the urban fabric, usually near residential areas, and ideally at a short distance from where children live. High backscattering, represented by very bright tones, enhances walls and fences exposed to the incident radar beam. The tallest buildings may be identified due to radar shadow related dark tones. A large lot is comprised of a parking lot, a high school (5-6,3), an ice arena (4,2-high), sport fields (5-6,4), and public services (5-6,2). A high-intensity residential area [(7-8,1-2) & (8,4-5)] at short distance from the school displays a bright tone and grid pattern. Intermediate tone patches represent large parking lots [(4,2-low) & (10,5)]. A dark tone and intermediate texture wide linear object is a narrow river alongside of roads [(2,2) to (5,5)]. A six-lane-divided road [(1,1) to (10,1)] displays a straight very dark tone line separated by a light tone median strip, and intermittent bright dots for lamp posts and traffic lights. Image: CSK 909 HH 18 deg RA Ottawa W 900 m LL 45.435384 -75.723805 deg.

incidence. In addition, the azimuth changing terrain locally affects the beam look direction. As a result, terrain vertical and horizontal shapes are interpreted by considering the system incidence angle and look direction. This

FIGURE 5.25
At the northern limit of Molinella, Italy, three buildings (5-6,3), which are in very close proximity to each other and aligned next to a railroad [(1,1) to (9,5)], are storage facilities of a horticultural cooperative. There is little equipment outdoors and limited parking space. The warehouse roofs display a dark tone, outlined by their south and east exposed walls that display fine bright tone lines. Except for the parking lots, immediately surrounding objects are intermediate and bright tones. The service industry area, along with a nearby shopping mall and residential areas, is composed of a high-intensity developed land cover type. The three land-use functions have distinct textures and patterns: a blocky pattern and coarse texture for service and commercial areas (3-4,1), and gridded for the dwelling area (6-7,1). And all of these contrast with the fine texture of agricultural dedicated open lands [(1-2,3-4) & (8-10,1-3)]. Image: CSK 806 HH 27 deg RD Ferrara W 2.0 km LL 44.628206 11.673900 deg.

section brings attention to landforms such as cliffs, hills, mountains, plains, terraces, and river valleys, some of which introduce strong contrasts, and others subtle image feature.

5.5.1   Cliff

A cliff is shaped as a sudden break of the terrain slope combined with an important elevation change. With radar beam look direction nearly perpendicular to a cliff, image foreshortening occurs and the slope displays a very bright tone over a line shape that fades on the ends due to change of the azimuthal direction, or the vertical slope (Figure 5.34). Covered by vegetation, the tone may be dimmed due to diffuse backscattering, but still well

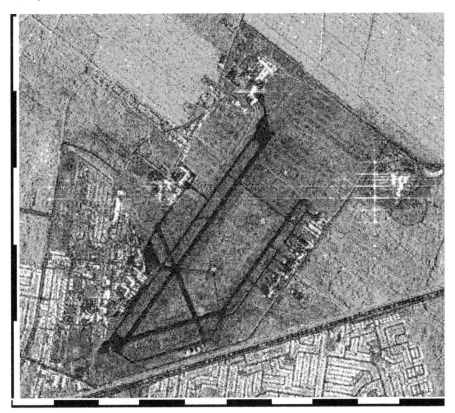

FIGURE 5.26
The 2,380 m main runway [(3,1) to (6,4)] of the Montréal Saint-Hubert Longueuil Airport can serve most regional aircraft. All runways and transit strips are sharply enhanced by the asphalt's very dark and fine texture that contrast with all other neighboring land use and cover types. A buffer corridor immediately surrounds the runways; it is either grass-covered or barren grounds cleared of tall standing vegetation and objects. In fact, a portion of it is farmed (7-8,4). Buildings on the property are low, except for the control tower. Clusters of rectangular and square buildings that house different operational functions cause very bright tone on the radar beam exposed sides. They spread on the property periphery next to the smaller runways [(3,2), (4-right,3), (6,5), & (7,2)]. Serving smaller aircraft, airfields may be found isolated from urban areas and surrounded with only a few buildings, but they will usually have a geometric layout and noticeable central runway. A succession of very bright star shapes is aligned with the location of a large antenna dish east of the airfield (9,3), on the John H. Chapman Space Center site. Image: RS2 476 HH 35 deg RD Montréal W 3.8 km LL 45.520235 -73.411115 deg.

FIGURE 5.27
At the interface of a water body and land, wharves are built as part of a harbor. The infrastructure has different dimensions, capacities, and functions that serve the transportation of goods and the mooring of industrial ships, ferries, fishing boats, or recreational boats. The Montréal harbor front of the Mercier-Hochelaga-Maisonneuve borough is a suite of piers, silos (5,3), container storage yards, and sharp bright tone producing cranes [(2-left,4), (5,4), & (7,3)]. Such a collection of equipment creates a coarse texture and irregular blocky pattern, with an overall arrangement in a linear fashion along the Saint Laurent River. The buildings and equipment, as well as a ship near the shore (8,3), display some of the brightest tones on the east side, i.e., up, that is directly exposed to the incident radar beam. The high-intensity industrial land cover expands over a width of 200 to 500 m. It is separated by a street, a dark line, and rail-tracks, a bright line, from a mix of high-intensity developments, including a military training facility (5,2), large buildings initially built to host transformation industries (3,2-3), high-intensity dwellings [(1,2), (6-right,1), & (9-10,1)], a quarry (8,1), and open land parcels [(1-left, 1-high) & (10-right,1-high)]. Image: RS2 088 VH 44 deg RD Montréal W 5.0 km LL 45.572232 -73.517777 deg. Note: Image rotated for display; north is to the left.

differentiated from the surroundings (Figure 5.4: (3,3)), mainly by the linear shape and gradually fainted contrast at the ends.

5.5.2 Hill

Hills are moderate height and gentle to moderate slope grounds. They may be identified by two distinct radar return strengths. A sharp bright tone is formed where the angular relationship of the terrain slope with the radar beam causes foreshortening, on the object's near range. Conversely, dark tones result from radar shadow on the terrain side that is not exposed to the incident beam. Shadow may be formed by trees on an open forest-covered hillside. In this

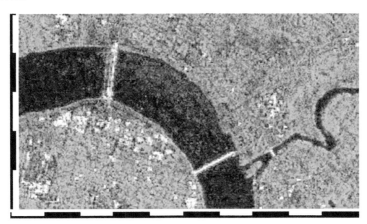

FIGURE 5.28
Bridges often cause very bright returns on radar imagery because of multiple backscattering on the metallic and angular constructions. In particular, railroad bridges have dense metal-meshed arches and guardrails. Sometimes, the structure and height cause a bridge to be represented by three side-by-side lines (3,4) due to distinct groups of reflection from the bridge (Raney 1998). The length of this linear object is determined by the width of the river or another transportation infrastructure above which it arches. To achieve the shortest path, the bridge crossing is usually at right angles to the obstruction. A bridge is best contrasted while it runs over calm water which has very dark tones. The road or train track leading to a bridge is normally perceptible. Additionally, given the width, which is likely to be wider for a multilane road than for a rail, the transportation purpose may be interpreted. Conversely, in some cases, the presence of a bridge is easier to notice and strengthens the interpretation of transportation axes (Figure 5.2: (5,3)). The illustration includes different bridge sizes and orientations at Fredericton. Crossing the Saint John River, the Westmorland Bridge (2-3,4) supports a two-lane street and the Bill Thorpe Bridge (6-7,2) is for pedestrians. To the east, the Route 105 Bridge (8-left,1) deck and Gibson Trail (8-right,2) hiking overpass cross Nashwaak River. Both the walking bridges are wood-resurfaced former railway bridges. Image: RS2 087 Four-polarization PCA 28 deg RD Fredericton W 4.4 km LL 45.962012 -66.631812 deg.

case, small dark spots contribute to overall lowering of the tone without forming a well defined very dark toned object typical to radar shadows. Otherwise, sharp radar shadow can outline an abrupt back slope in a contrasted manner similar, but opposite to that produced by foreshortening (Figure 5.35). Due to the gentler slope and rounded shapes of long vertical and horizontal dimensions, hillsides bright tones are displayed over wider linear shapes and are less contrasted than for a cliff.

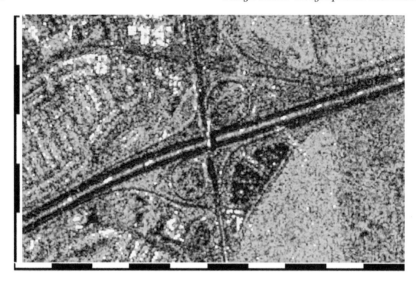

FIGURE 5.29
The interchange of the Trans Canada Highway and Eagleson Road, Ottawa, is
a trumpet intersection (Leish and Morrall 2014). At that location, motorists
can go between eight- and four-lane-divided roads. Specular backscattering on
the asphalted surface is represented by dark tones, which are further enhanced
where a lane is wider and is aligned with the radar beam right ascending look
direction. The concentration of large dimensional curvilinear shapes stand
out among most land covers. The highway median displays intermediate tone
and regularly distributed bright dots representing lamp posts. An interchange
right-of-way is relatively wide, but the external ramps may closely border other
developed lands. Nearest to the infrastructure are landscaped lands (4,3),
commuting park-and-go facilities (7,2), and commercial services (4-5,1). Inside
of road loops and on the road sides, windbreak vegetation is maintained [(5,2)
& (6,4)]. These are characterized by a coarse texture and dark to intermediate
tone. Farther around the interchange, land is developed for public service
(1-2,5) and high-intensity residential developments (1-3,3-5), while crop fields
cover the lands to the east [(7-10,5) & (8-10, 1-3)]. Image: CSK 909 HH 18
deg RA Ottawa W 1.3 km LL 45.320436 -75.886083 deg.

5.5.3   Mountain

Mountain ridges expand linearly over several kilometers and connect at high
altitude to form much larger mountain ranges. Crests stand out in the landscape by a sequence of bright and dark tones for the exposed and back slopes,
respectively. In addition to the foreshortening, this strong topography is prone
to creating image layover. The brightest tones correspond to only a portion
of the exposed slope at the location of a perpendicular incidence of the radar

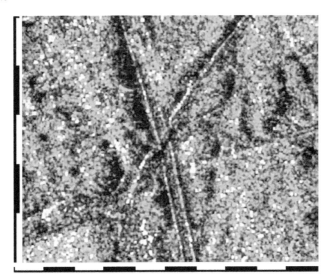

FIGURE 5.30

The radar shadow helps identify a road overpass. It shows as a short dark line on the upper-level transport axis far range side. The shadow is slightly longer than the underneath axis width, as it also arches over the road side slopes. South of Manotick, Ontario, the two-lane Prince of Wales Drive overpass at the four-lane-divided Veterans Memorial Highway displays a sharp dark line, slightly offset to the southeast (5-6,3). This intersection is a simple crossing, with no exit ramp, and it is surrounded by uneven height forest patches. Some of the bright tone linear features along the road sides [(4-right,2) & (6,4)] may be explained by foreshortening where the radar beam is incident on ditch slopes and metallic guardrails. The road surfaces on this image do not cause a typical specular backscattering and therefore are not particularly dark in tone. This is likely due to the lanes being narrow and at an azimuthal orientation of 45 to 90 deg from the radar beam look direction. In this case, the radar beam may not reach the road surface, or a specular backscattering may not follow its course away from the surface, because it interacts with higher forest stands nearby. Image: CSK 909 HH 18 deg RA Ottawa W 1.0 km LL 45.194298 -75.724322 deg.

beam. Image layover which occurs where the radar beam facing terrain slope is larger than its incidence angle is more difficult to visualize. It often gives the appearance of pointed shapes, with their apexes oriented toward the radar antenna (Figure 5.36). Tall objects on a back slope may have their height exaggerated by radar shadow from a lower radar beam grazing angle, i.e., very large local incidence angle. Due to their elongated shape, mountain ridges and ranges appearance on radar images are sensitive to the beam angular

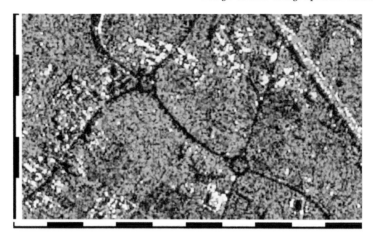

FIGURE 5.31

Roundabouts are ring shaped intersections with three or more outwardly connected roads. The asphalted surfaces, which are at least two lanes wide, have a dark tone stand out due to specular backscattering, particularly if the road sides are kept clear of vegetation or other objects. The contrast is excellent as there is no obstruction to the radar beam. At the center, a large roundabout may be landscaped with vegetation, decorative infrastructures, monuments, or even buildings. The image represents roundabouts of routes SP 31 (4,4) and SP 32 (7,2) intersecting with the Via Ferruccio Parri by-pass road south of Cortona. While the SP32 is a little narrower, all are two-lane roads crossing agricultural industry-dedicated low-intensity developments. The roundabout centers are dark toned and intermediate textured lawns, surrounded by posts that remain unnoticed on the image. Herbaceous planted land cover dominates the surroundings while trees, isolated (9-left,3-high), and in rows [(5,2), (2,4-low), & (10,3) to (8,5)], are shadow enhanced. Tree rows may look similar to roads by their linear shape and dark tone, but roads have a steadier width, and are associated with intersections and are alongside buildings. Image: CSK 212 HH 31 deg RD Cortona W 1.1 km LL 43.254434 11.970288 deg.

variables, including the incidence angle, look direction, and orbital path orientation. Coverage based on several options of these system specifications must be envisaged for creating spatially continuous LULC interpretation of mountainous landscapes.

5.5.4 Plains

Plains are sediment deposited environments expanding on very gentle slopes, whose length is several meters to kilometers long with various widths or discontinuities, at the margins of oceans, lakes, and rivers, current or ancient.

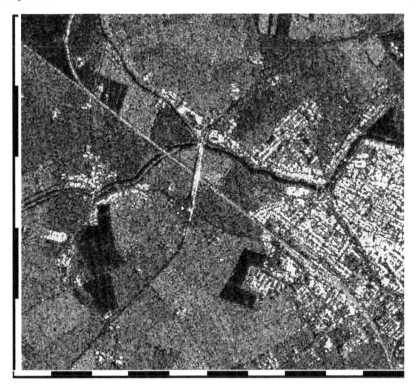

FIGURE 5.32

Railroads are constructed of metal tracks and spikes, in addition to an infinite suite of crossties. This structure is corner and diffuse backscattering prone, which creates a bright tone overall. Where the rail is near-perpendicular to the radar beam look direction, the tone is much brighter. Skirting southeast of Portomaggiore, a two-lane railroad displays a relatively bright tone as a narrow line shape [(1,5) to (10,1)]. At a location where it is oriented perpendicular to the radar beam (2,5), the tone is very bright. Oblivious to most landscapes, the path shape is straight, and wide-radius curves allow direction changes. A long road bridge (5,3) crosses over a canal, the railroad, and a street. At the south boundary of Portomaggiore, the train station is identified by several tracks forming a wider linear object (8-9,2). As it reaches urban areas, harbor fronts or airport grounds, a single track is replicated by many side-by-side segments and bordered by buildings to facilitate traveler commuting, triage, storage, and transit of goods. Image: CSK 806 HH 27 deg RD Ferrara W 2.3 km LL 44.701196 11.791070 deg.

Conversely to cliffs, hills and mountains, the nearly flat ground of plains gives way to the land cover to control the representation of tone, texture and other image interpretation criteria. If it is bordering a water body, plains are subject to intermittent flooding and only a small water level rise is needed for a

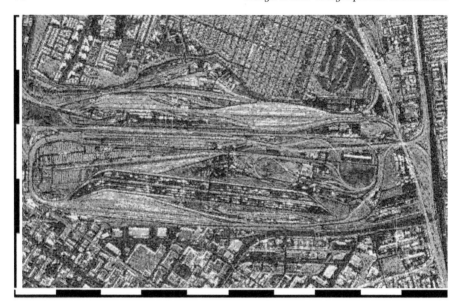

FIGURE 5.33
A large yard in Montréal connects to national and international railways.
It is in proximity to airports, harbors, and highways. It is a major facility
connected to other shipment points in Canada and the United States. The
site is covered by numerous rails, onto and between which containers and
wagons are stored. Side-by-side rails and other functions for the yard are
represented by dark to bright tones, and variable textures. Wrapping around
and grouped tracks display some of the brightest tones in a striated pattern
[(3,2), (4,3-low), & (6-7,4-low)]. Occupied parking lots [(2,3) & (8-right,4-low)]
display short bright lines in a grid pattern, which is contrasted on the dark
tone asphalt background. A coarse texture that extends over a narrow band is
produced where containers lay, somewhat scattered, on an asphalted surface
[(3,2-high) to (7,2-high)]. Curved rails on the north and south ends define the
property extents. Just south, i.e., right side, the very-dark tone fine texture
wide line is the Trans Canada Highway (10,1-5). High-intensity developments
for service industry (1-10,1) and for residential use (1-7,5) border the area. An
open forest land (10,2-3) separates the highway from the rail yard; and a golf
course (8,4-5) is tucked by an east turning track. Image: RS2 088 VH 44 deg
RD Montréal W 4.4 km LL 45.466784 -73.684982 deg. Note: Image rotated
for display; north is to the left.

large territory to be inundated, but not necessarily immersed. In this case, the
land cover becomes a mix of water and vegetation (Figure 5.37), and, in some
cases, buildings. As a result, the radar backscattering on emerging objects is
very high because of double and multiple radar beam bounces that are created
when they are surrounded by water.

FIGURE 5.34

An abrupt slope of Kaneshekat Mountain is on the Grand Lake south shore [(3,4) to (7,2)], Labrador. Over the steepest sections, at the location of brightest tones (5,3-high), the slope reaches 85%, or 40 deg. The 2.5 km long feature emphasizes the northeast facing horizontal cliff drop, parallel to the Grand Lake shoreline. With the foreshortening related bright tone, layover may occur where the slope exceeds 37 deg, which is the radar beam incident angle for this image. Treed vegetation covers the cliff, except for some openings near the summit. Intermediate texture and mottled pattern characterize the forest land (1-5,1-2) and water surface (6-10,4-5), but in different ways by the tone that is overall brighter and the patches less contrasted for the forest. Image: RS2 475 HH 24 deg RD Labrador W 3.5 km LL 53.641771 -60.434867 deg.

5.5.5   Valley

Brooks, streams, and rivers are entrenched in river valleys. These relatively narrow depressions may host water intermittently or permanently. In a well-developed network, the downhill end of a stream is connected to a higher hierarchic level segment (Strahler 1952) for eventual linking to a lake or the ocean. Structural breaks, such as fractures and faults, variable rock types, and surface deposits, contribute to create water channels. On synthetic aperture radar images, well-entrenched tributaries are often enhanced by the valley hillsides, not necessarily by the water flow. A drainage system provides information on the relative terrain elevation and the extent of a watershed. From the low to high hierarchic stream level, the terrain altitude decreases. Steep valley sides that are exposed to the radar beam cause bright tones, as opposed to a radar shadow that marks a slope turned away from the incident radar beam (Figure 5.38). In pairs these are parallel lines, especially for v-shaped valleys. The effect is greater when a valley is oriented perpendicularly to the radar beam look direction. The bright-to-dark sequence is from

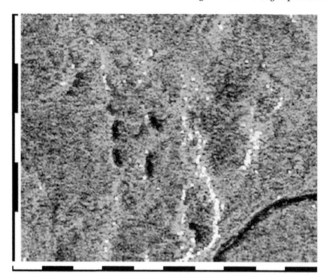

FIGURE 5.35
The imaged eastern portion encompasses a relatively level Goose River valley
(9-10,1-2) area, with the terrain elevation increasing in steps toward the west,
in the Boreal Shield direction. The terrain height near Goose River is about
15 m and it reaches to just above 300 m to the west. Hill sides enhanced by
a bright tone on east and southeast facing slopes [(1-2,4), (4,1-2) to (3-right,
4-5), (7,1-3), & (8,3)] are near-perpendicular to the radar beam incident from a
right look descending orbit. Dark tones and a coarse texture reveal the hills far
range-side rounded shapes [(1-2,5), (4-6,1) & (8-left,3)]. The darkest tones on
the image are water bodies, including lakes (4,3) and Goose River, and radar
shadow delimiting forest patches next to logging lands [(2,1) & (8,4)]. Image:
RS2 475 HH 24 deg RD Labrador W 5.4 km LL 53.388641 -60.552482 deg.

far to near image range, representing the exposed and opposed side of a valley oriented perpendicularly to the radar beam. This sequence is reversed for
hill ridges, where a bright-to-dark suite is from near to far range. This feature helps differentiating topographical features and other objects rising above
ground from these shaping a depression.

5.6   Recreational

Recreational land use generally constitutes high or medium intensity developed land cover near residences and schools, or part of a service industry hub

FIGURE 5.36

North of Cortona, Toppo di Morro is located (3-4,3) in the rolling highlands of the Apennines. It is surrounded by large northeast to southwest oriented ridges [(1,1) to (4,5), and (6,1) to (7,5)], and short secondary northwest to southeast crests [(2,3) to (4,1)]. Throughout, many abrupt slopes characterize this complex terrain topography. Dark tones denote dense forest canopies that lie on far image range oriented hillsides [(8-right,4-5), (7,1-2), & (2-right,2)], roadside forest edges (1-2,5), a small pond (4-right,1) and cleared land [(9-left,3) and (10,3)]. In mountainous areas, the local incidence angle changes from place to place, and takes precedence in setting the image tone. Contrasting dark and bright tones normally brought by vertical objects is lesser on a radar beam exposed slope, to the point of foreshortening occurring, where land cover types are all represented by a very bright tone. This condition is shown by the bright curves and v-shape lines (8,1-2). Image: CSK 212 HH 31 deg RD Cortona W 3.2 km LL 43.332604 12.006535 deg.

linked by transportation axes for access by the public. Buildings are specifically designed to address particular requirements of one or few activities. Outdoor sports and recreational infrastructures are generally planned over large expanses of natural ground and include very few buildings. Thus recreational land use type often fits open space as a land cover category. The ensembles presented in this section stand out by pattern, shape, and dimensions. Several key components enable their identification. Examples include arenas, a golf course, hippodrome, marina, ski resort, and outdoor sport fields.

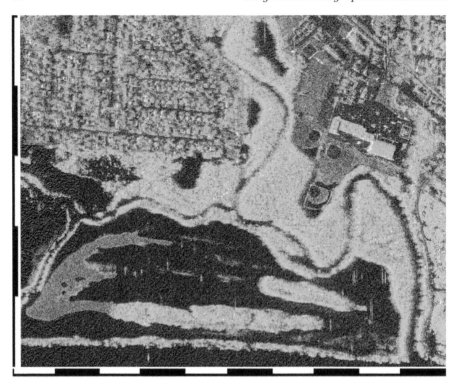

FIGURE 5.37
The image was acquired on a day the water level had risen by about 4 m,
inundating the Nashwaaksis River plain at the junction with Saint John River.
Bright tone intermediate texture is that of emerging deciduous forest canopies
(7-left,3), and dark tone fine texture is partially immersed grassy fields (2,2-
low). The very dark tone shows open water in the normal Nashwaaksis Stream
(10,1-2) and Saint John River (1-8,1-low), the flooded plain (4-5,2-low), and
intermittent ponds [(6-right,5) & (6,4)]. Where grass and agricultural lands
are completely inundated, specular backscattering produces a very dark image
tone, but a few bright spikes indicate emerging grasses (4-5,2) and tree rows
[(1-9,1) & diagonal at (1-2,2)]. Spared land, which is just over the 10 m ASL
height, includes high-intensity developments residential (4,4), commercial services (9,4), and sport fields (8,3). Image: CSK 936 HH 41 deg RA Fredericton
W 1.3 km LL 45.975938 -66.661009 deg.

5.6.1   Arena

Indoor ice-skating facilities with limited seating take up a few hundred square
meter footprint in the landscape. Constructed usually following a simple archi-
tecture, rectangular shape, it is dimensioned to suit an ice surface and a
few bleachers. On a gently sloped roof top, the backscattering is dominantly

FIGURE 5.38
A tributary of the Nashwaak River (1,1-5), just south of Taymouth, the McBean Brook and upper streams form a dendritic system that cuts through a moraine blanket and fluvioglacial sediments (Rampton 1984). The terrain height from Nashwaak River, at the tributary, to the lower hierarchic level streams (10,2), increases from about 25 to 180 m ASL. Some of the smaller streams upland are better demarcated as their orientation is directly facing the right looking beam from an ascending orbit. A dense forest covers most of the area, producing intermediate tone and texture. A subtle striped and block pattern indicates the land is logged (4-7,1). A transmission line runs through the landscape, perpendicular to McBean Brook (3,1-5) and abrupt slopes lead to the Nashwaak River valley. One of them, a 30 m altitude drop, is marked by a very bright curvy line (2-left,1). Image: CSK 953 VV 41deg RA Fredericton W 8.2 km LL 46.167966 -66.567265 deg.

specular. Contrasts occur due to the bright tone from the radar beam foreshortening or corner backscattering from the exposed building wall, refrigeration equipment, and rooftop ventilation outlets. Radar shadow on the far image range side exhibits a low building height (Figure 5.39). Considered for various indoor sports, large audiences, and entertainment facility, multipurpose arenas are designed as very large buildings, of moderate height, and circular, oblong, or rounded-corner-rectangular in shape (Figure 5.40). Such large arenas are located to facilitate access by the public, near large road arteries, surface passenger trains and subway lines. Modern arena projects at the outskirts of high-intensity developments draw on land available for parking areas and complementary sport facilities, and service industries.

5.6.2   Golf Course

Golf courses exhibit a ribbed pattern of alternating dark and medium tones due to the grass covered fairways, separated by trees. Bright tones may represent a clubhouse and vehicle storage buildings. The overall shape of golf

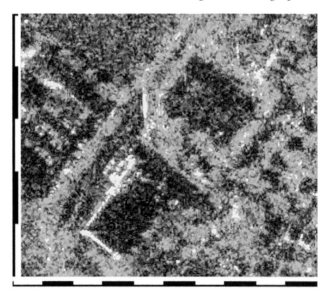

FIGURE 5.39

The Lady Beaverbrook Arena (4-5,1-2) is grouped with outdoor sport fields in the heart of a mature medium intensity development of dwellings (8-9,2) and apartment buildings (1-2,3-4). The ice arena building is rectangular, a little larger than the standard 60 m long ice surface it houses. A dark tone and fine texture differs from parking lots that service the sport facility. They also have relatively smooth surfaces, but display a dark tone and mottled texture [(6,2) & (7-8,4)]. The building is enhanced by foreshortening on two radar beam exposed sides, while the parking lots show no evidence of a height dimension. Deciduous trees in rows [(9,4) to (7,3)] and forest patches (2,1) border the arena and parking lots. As the image was acquired at the end of a winter season, volume backscattering on leafless branches and trunks display intermediate tone and texture. Many trees are enhanced by their radar shadow and canopy openings on barren lands and lawns display dark tones and rounded shapes [(9-3) & (3,4-high)]. Image: CSK 936 HH 41 deg RA Fredericton W 300 m LL 45.951318 -66.638358 deg.

courses varies from oblong, square, rectangle, or linear open spaces. Fairways may be molded by natural barriers, recreational attractions, or integrated to the urban fabric alongside other land use types (Figure 5.41).

5.6.3   Hippodrome

A hippodrome is a singular recreational infrastructure. It displays a prominent rectangle tightly enclosing a dark tone oval, street-width race track. The grass

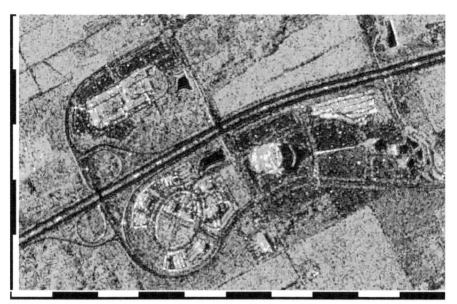

FIGURE 5.40

In Kanata, Ontario, a 150 m diameter building was constructed to host National Hockey League (6-7,3-low) games and multiple other sport and cultural events. The building roof produces intermediate to bright tones, coarse texture, and a few linear-shaped very bright features. With a seating capacity of 17,000, this type of facility requires ancillary commercial and restaurant businesses, and much parking space (8,2-3). Neighbors are car dealerships (4-5,1-2), a shopping mall (2-4,4), and PST service industries (9,3). Radar shadow, by the building side toward the image far range, yields height information: in increasing order, the mall, a single building to the south (6-right,1), office buildings in a three-set, and the arena. Ponds in the area display regular shapes that suggest they were configured to help manage surface waters. A few [(10,3) & (9,1)] are linked to a small stream. Agricultural lands and forest patches [(6,1) & (1-3,1)] surround the sport and service hub, which is accessed by a four-lane divided highway. Image: CSK 909 HH 18 deg RA Ottawa W 2.1 km LL 45.296352 -75.929878 deg.

covered center and borders are a little contrasted by their intermediate tone. Sparse equipment and trees may show bright tones. As well, spectator stands may take up a considerable part of the whole infrastructure on the sides of an oval. Horses and racing gear require stables and other large buildings (Figure 5.42).

FIGURE 5.41

The Ottawa Hunt and Golf Club 18-hole course displays dark and intermediate tones representing the fairways and trees, respectively. The area is compact and about rectangular in shape [(2,1) to (9,3-4)]. Small very dark tone features suggest ponds [(3,2) & (8,4)]. The club house and parking facilities are accessed by the North Bowesville Road (4,2). The golf course surroundings include high-intensity residential developments [(2,4) to (8,5), & (10,2-5)], professional-scientific-technical service industries [(3,5), (2-3,1), & (6-7,1)], grass and shrub lands [(1,2) & (8-9,2)], and a deciduous forest (8-9,1). Image: CSK 032 HH 18 deg RA Ottawa W 1.9 km LL 45.342604 -75.682509 deg.

5.6.4   Marina

Recreational boating requires harbor facilities that minimally consist of a docking space, slipway, and some parking area. The overall expanse for a marina depends on the number and size of boats it accommodates. Very large yachts, cruise ships, and ferries also use industrial-type water fronts that have more space between piers or that offer a simple docking area aligned with the coastline. A marina may be open or sheltered from the water body it is accessible to. Embraced by long piers or wave breakers, marinas change natural shores into rigid shapes (Figure 5.43).

5.6.5   Ski Resort

Downhill ski resorts have a peculiar design in the landscape. They consist of open space land cover, of a few mechanical infrastructures and buildings. Ski trails show as narrow corridors cleared of forest that converge at their

FIGURE 5.42

The Fredericton Capital Exhibit Centre hosts harness racing and agricultural shows. The 800 m race track is the central infrastructure. It displays a very dark tone oval narrow linear shape, and dark tone medium texture for the inside field. Various size buildings [(5-6,1) to (8,4)] and a large parking lot (8-9,2) are just nearby. Both display a dark tone and fine image textures, but the building differentiation is from the narrow bright line that outlines the radar beam exposed walls. A few bright spots for cars and light posts are scattered in the parking lot. To the southwest, a set of long sloping metal-roofed buildings would have caused the very strong backscattering (2-4,1), given the radar beam is right looking from an ascending orbit. Farther off the area, land use functions include low (1-2,4-5) and high (10-1) intensity residential. Immediately surrounding the hippodrome are outdoor sport fields (7-10,5), services industries (3,3), and a curling club house (2,2). Image: CSK 936 HH 41 deg RA Fredericton W 500 m LL 45.960979 -66.658417 deg.

ends, and are parallel or wide apart midway. Foreshortening and radar shadow enhance the forest edges along the trails. This configuration has the natural forest or remaining tree rows work as wind breaks for the cleared trails (Figure 5.44). However, ski resorts may be designed in tree-less environments. In this case, the bright tones locating gondolas, chairlifts, communication towers and building clusters may lead to their identification. Further, road access and

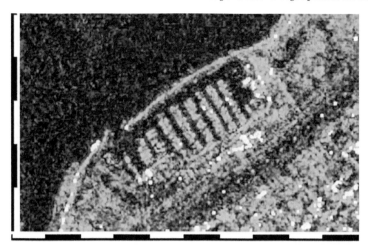

FIGURE 5.43
Réal-Bouvier Marina is on the Saint Lawrence River shore at Longueuil, Québec. Besides the water surface (1-3,4-5), dark tones represent an asphalted parking lot (5,1-high), and a four-lane-divided highway [(5,1) to (10,3)]. A dozen piers to accommodate 400 docking places (SOGERIVE 2006) is sheltered by a long tree-covered breakwater [(4,3) to (6,4)]. A narrow passage (4-left,3) to the river aligns with the launching and berthing lift. This arc metallic infrastructure displays as a large point shape bright feature on the image (5-left,2-low). Likewise indicated by a bright tone cluster (7,3-low) is a building that houses restaurants and professional services. Lamp posts produce small bright dots aligned on the highway median and cars parked near the clubhouse (6,2-low) would create the bright coarse texture, while the image was acquired on a Sunday in July. Image: CSK 433 HH 22 deg RA Montréal W 900 m LL 45.541103 -73.514702 deg.

high density developments are often near the base of ski resorts. The trails may display a finer texture than the surrounding forest cover. A whole resort expands over several hectares, on mountain sides which are preferably North oriented for least sun exposure on the snow cover, which prolongs the ski season.

5.6.6   Sport Fields

An outdoor sport field essentially consists of a determined size (STMA 2018) playing surface and a few, or no, contiguous buildings. Playing surfaces are made of grass, barren soil, artificial composite, or a combination thereof. The level terrain displays a very dark or dark tone and smooth image texture.

FIGURE 5.44

Camp Fortune Chelsea, Québec, North of Gatineau, is a ski resort during winter, and aerial park at summer time. Two dozen runs and seven lifts are in three groups [(2-3,1-2), (6,3), & (7-8,2)]. Bright single points mark lift base locations [(7,4) & (9,2)], and larger bright rectangular shapes (4,2) are at the location of ski lodges. From there, some of the lift lines and pylons are noticeable by faint dot-marked bright lines [(2-4,2-low) & (7-9,2-low)]. Other linear features on the image include roads and a power transmission corridor. A narrow winding road (5-6,5) connects lodges to the mountain base roads (6-7,5). The narrow power transmission corridor differentiates from the other linear features by its straight path across the hills (6,4). The overall landscape is forest covered. Gradual variations of dark and intermediate tones reveal a moderate topography. A few sharp slopes are enhanced by darker [(10,1) & (8,4)] and brighter [(5,3), (3,4), & (9,4)] tones. Image: CSK 909 HH 18 deg RA Ottawa W 2.4 km LL 45.513744 -75.845367 deg.

The standard shape and dimensions of sport fields are reliable keys. In a developed land cover environment, indoor and outdoor facilities may be grouped on lands where other services are offered by municipalities, schools and universities, which are the organizations that often promote sports (Figure 5.45).

FIGURE 5.45

A five-hectare block in the city of Fredericton contains two softball diamonds (2-6,3-4), a multipurpose field (7-9,2) sized for American football, rugby and soccer, an area (7,3-4) with a basketball and three tennis courts, and a swimming pool (5,4). Particularly, the baseball diamonds have an identifiable shape, as well as a distinction between the square- or circle-shaped dark tone of the smoothly surfaced infield compared to the grass-covered outfield. Represented by very fine bright lines, fences or mesh barriers appear to border portions of the infield and outfield. Also, high metallic fences delineate the pool and tennis court areas. Image: CSK 936 HH 41 deg RA Fredericton W 400 m LL 45.953217 -66.640487 deg.

5.7 Vegetation

Vegetation presents a wide range of tones and textures on a synthetic aperture radar image, so information may be obtained about the type, vertical structure, canopy density, and other characteristics such as moisture, exposed understory and supporting terrain morphology. This section illustrates vegetation covers in different contexts including forests, fire altered forests, natural and planted herbaceous, grasses, single trees, and tree rows. Previous rubrics have discussed forests in a state of flooding, altered by industrial logging and various topographies.

5.7.1   Forest

Forests occupy areas from confined to extremely vast. They display inter-mediate or bright tone and coarse image textures. Leafless deciduous trees enable volume backscattering, which results in a higher image tone than for coniferous and leaf-on deciduous stands (Ahern, Leckie, and Drieman 1993). Conversely, surface backscattering on wide leaf-on tree crowns appears darker than deciduous canopies. Natural forests have consistent appearance through-out with soft tone and texture transitions. The surface canopy, and understory, leaf and wood components, interacts through surface and volume backscatter-ing, respectively. Dark tones and fine image textures may be fashioned by a dominating surface backscattering representing very dense forests developed on even terrain. The variation of features such as tree species, stand height, opening canopy, topography, age, and vegetation or grounds of the understory exposed to the radar beam might increase image brightness and coarseness. Discontinuities may also be noticed at the location of slopes or streams but human related alteration can be observed due to the emergence of abrupt boundaries defined by tones, image textures, and patterned areas. Bois Pap-ineau (Figure 5.46) is located Southeast of the 440 and 19 Routes intersection, in Laval, Québec. It is a dense beech tree dominated forest on most of a 100 ha area (ACPB 2017).

5.7.2   Forest Fire Scar

Post-fire forest regeneration success depends on the climate, substrate, micro-topography, moisture of the ground, and other factors that facilitate seedling and growth (Abbott, Leblon, Staples, Maclean, and Alexander 2007). Scars on the landscape may be evident for many years and cover several hectares. Burns have various shapes and dimensions resultant from the fire dynamics, natural barriers and firefighting efforts. Recent burns expose debris, soil, and rock. Tree species succession over the years is uneven compared to the surrounding undamaged forest stands. During the spread of a fire, the wind direction may govern linearly-shaped features, which persist throughout the canopy regrowth (Figure 5.47).

5.7.3   Herbaceous, Natural

Natural herbaceous, similarly to cultivated, is woodless. Consequently, the tone appears dark to intermediate, and the image texture is much finer than that of forests. Overall, homogeneous tone and textures may give room to gradual changes related to grass density, species variety, moisture state, or terrain shape. The border of herbaceous vegetation fields by short grass or treed lands are often outlined by contrasted tones of corner backscattering, foreshortening or radar shadow. Located in Ottawa the Mer Bleue Bog raised

FIGURE 5.46
The fine texture and intermediate tone of the forest [(3,2) & (5,3-4)] are in contrast with the urban environment, including high-intensity residential developments [(1,3-4) & (9-10,3-4)], and the cloverleaf intersection (3,3) of four-lane divided highways. Subtle variation of tones in this forest is attributed to variable growth stages and densities. The understory is primarily vegetated and trails are in place for encouraging all-year round outdoor activities. Linearly shaped darker tones trace a wide trail (6,4-5), where a larger specular backscattering component occurs on short vegetation and barren grounds, and a northwest facing terrace [(6,1) & (8,3)] is radar shadow enhanced. A bright straight line highlights a single track railroad [(4,1) to (8,5)] and the Ruisseau La Pinière stream shore alongside the hiking path. Image: RS2 VH 44 deg RD Montréal W 2.9 km LL 45.603325 -73.689003 deg.

wetland (Figure 5.48) is principally natural herbaceous vegetation covered (Ramsar 2001). This land cover slowly changes to forest canopies at a few meter higher altitude locations of this dome shape unit. The image represents about a tenth of the total Mer Bleue Bog area. The brightest tones within the raised wetland area represent marshes, the intermediate tones are the treed bogs and upland forests, and darks correspond to sedge-, fen- and shrub-dominated bogs (Baghdadi, Bernier, Gauthier, and Neeson 2001; Touzi, Deschamps, and Rother 2007).

5.7.4 Herbaceous, Planted

Planted herbaceous is a land cover description that refers to woodless plants. It generally refers to a large category of crops in the agriculture industry which differentiates from permanent crops such as vineyards, olive, and fruit trees. Factors such as the crop type, growth stage, and moisture produce many different backscattering intensities and consequently a large range of

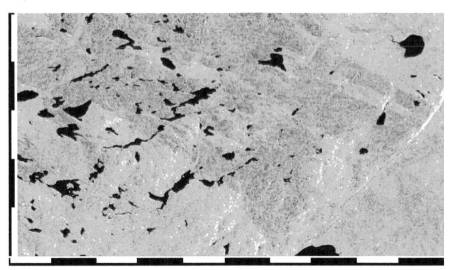

FIGURE 5.47

Forest fire scars are noticeable on a portion of a 23,000 ha forest burn north of Happy Valley-Goose Bay. During the last week of June and the first two weeks of July 1985, the fire progressed toward the southeast. The image captures the 23rd year reforestation, which is estimated half-time period for a forest of this type to reach an undisturbed stage (Miranda, Sturtevant, Schmelzer, Doyon, and Wolter 2016). Of similar image texture, intermediate tone, large rectangle-shaped burn scars are contrasted with the adjacent bright tone forests, and from the dark tones indicating wetlands and lakes. Some of the larger lakes [(2,4-low), (3-right,5) & (7-left,5-low)] appear to have acted as barriers to the fire progression as regrowth and mature forests are on their northwest and southeast sides, respectively. Some topography related bright features [(7,2) & (8,5)] are more evident in the burnt area because less treed vegetation dampens the contrasts. However, in a recent burn, large debris that are left uncovered may cause a coarse image texture or a speckled pattern. Image: RS2 474 HH 37 deg RD Labrador W 20 km LL 53.485031 -60.411415 deg.

tones (Bouman and Hoekman 1993, McNairn, Ellis, Van Der Sanden, Hirose, and Brown 2002). However, this land cover displays prominent spatial context characteristics. Throughout the different tones, the image texture remains fine and the pattern is geometric, regular or irregular. Stripe and dot patterns that emerge within lots indicate crops in row or plow marks on barren grounds. Field edges may be enhanced by dark or bright narrow lines, representing radar shadow and foreshortening caused by dividers such as tree or shrub rows, ditches, fences, drainage channels, stone walls, roads, trails, or terraces (Figure 5.49).

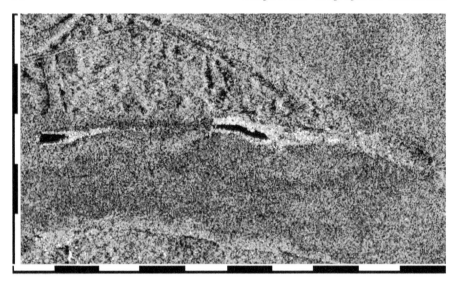

FIGURE 5.48

The herbaceous vegetation land cover tone is overall intermediate (1-6,2) and the texture is fine to intermediate. In sparsely treed large sections the tone is slightly brighter [(8,2) & (10,4)]. A transition buffer about 100 m wide is a shrub vegetation environment [(5,1) & (9,3)] that displays a bright tone and intermediate texture. The brightest of tones in the wetland represent marshes where cattails and alders are the dominant species, and they form narrow bands wrapping around small ponds [(2,3) & (7,3)]. On land about 10 m higher, outside the bog, a cover with alternating grass fields and forests create intermediate (3,1-low) and coarse (3,4) image textures, and a contrasted mottled pattern from radar shadow delimited tree patches (5,4). Image: CSK 909 HH 18 deg RA Ottawa W 2.2 km LL 45.392123 -75.510565 deg.

5.7.5   Lawn

As a sort of planted herbaceous land cover, grass lawns form open spaces in developed areas, to serve horticultural designs or topsoil protection. On private and public property lots, lawns are usually kept mowed and healthy. An even grass cover is made of very compact leaf plants, growing to up about 5 cm. With both the X-band (Table 2.1) and C-band (Table 2.2) synthetic aperture radar wavelengths, this corresponds to a rough surface texture, but as the plants are much closer to each other, there tend to be specular backscattering (Figure 5.50). Grass used for lawns, sport fields and road sides is represented on radar images by a dark tone and a fine image texture. Tones that are darker than lawns on synthetic aperture radar images are other much finer surfaces such as calm water, asphalt, and compacted barren surfaces. Radar shadow, although it is system design related, is also very dark on the image.

FIGURE 5.49

A region located in the Po River Valley, surrounding the Portonovo community population of 300 (1,1), displays a wide range of field dimensions and
tones, but an overall homogeneous texture. The tones convey a diversity of
crop types and growth stages. Clusters of buildings are scattered through the
landscape, approximately showing land ownership. On the west (1-3,1-5) and
east (10,1-5) image sides, the fields are large, while those in the center are
much smaller. This contrast may be related to factors such as the topography,
soil type, land-reclaiming history, and ownership. Except for a few fields where
a pattern of narrow stripes is observable [(1,5), (10-left,2-3), & (5,4)], intermediate textures characterize most of the region. Greater linear features in the
image include an alignment of electrical power transmission pylons (2,1-5),
streets, and a narrow channel (4,1-5) bordered by high herbs and shrub covered levees. Image: CSK 806 HH 27 deg RD Ferrara W 3.2 km LL 44.534742
11.772998 deg.

5.7.6   Tree Row

Trees in rows have practical applications such as property delimitation, environmental buffer zones for streams and landscape alterations by humans, wind
breaks, or for permanent crop designs and reforestation strategies. The tone
is bright on the image for tree crowns and row sides exposed to the radar
beam, where corner backscattering occurs, and dark due to a radar shadow.
The best context for identifying this contrast is where the long-axis dimension of a tree line is oriented perpendicularly to the radar beam look direction.
Narrow and tight tree lines convey sharp contrasts, while rounded crowns and

FIGURE 5.50
Nashwaakis, situated north of Fredericton, is a high and medium intensity development with mainly residential land use [(1-2,2) to (10,1-4)]. It borders a four-lane-divided highway [(1,3) to (10,5)] and large forested open spaces [(1,4), (3-4,5-high), & (5,3)]. Lawn areas are scattered throughout the developed land cover. They display a dark tone and a fine to intermediate texture. The tone is lighter than the widest roads and radar shadows (9-right,3), and the texture is finer than forest lands. Lawn shapes are usually rectangular, square (10-left,3-high), or irregular according to the lot dimensions [(3,3), (9-left,1), & (9,3)]. In residential areas, individual lots may not be perceptible but, overall, grass-covered yards display darker tones than treed yards. Also, lawns and houses, with no trees, reveal aligned bright dots and more tone contrast [(4-5,3) & (4-5,4)]. This feature may help differentiate residential development characteristics such as building height and proximity, and, sometimes given the tree heights, the neighborhood age. Image: CSK 906 HH 24 deg RD Fredericton W 2.6 km LL 45.988235 -66.656114 deg.

loosely aligned trees introduce diffuse backscattering from a larger volume of branches and trunks interacting with incident beam energy (Figure 5.51).

5.7.7   Tree, Single

Isolated trees are found to grow naturally or on landscaped yards. Leaf-on trees that have a wide top are better demarcated from their background compared to pointed top trees. The evidence of a tree is very small but follows a systematic pattern. A bright tone rounded shape object, representing diffuse backscattering on the tree crown, is just next to a radar shadow. The bright

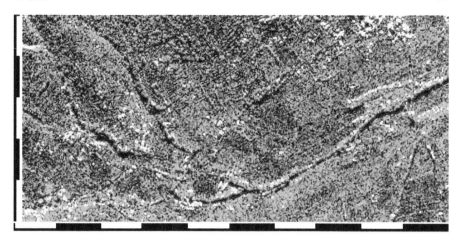

FIGURE 5.51
Nearly continuous tree rows marked by sometimes thick dark lines run across the landscape. With a favorable orientation on the tree rows' near-range side, foreshortening produces bright tones [(3,2), (6,1) & (8,2)]. The tree rows' intermediate texture is similar to that of forest patches [(9,3-high) & (6-right,4-low)] close by. Most of the permanent crops are developed in small tree-bordered fields (7-8,4-5). The crops display medium and coarse textures, some that spread along valley slopes exhibiting a faint dot (4,4) or striped (6,5) pattern, where the trees are distributed in rows. Image: CSK 212 HH 31 deg RD Cortona W 2.2 km LL 43.215475 12.065235 deg. Note: Image rotated for display; north is to the left.

and dark shapes, to the near and far image range, are mirrored. Also, the image texture is a little coarser for the bright tone than for the shadow area (Figure 5.52).

5.8 Water

Water bodies take many dimensions and shapes, and because they may cover very large expanses, their representation at a large scale is sometimes only partial. Calm surface water displays some of the darkest tones on synthetic aperture radar images due to specular backscattering of the radar beam. It presents as a dark tone and fine texture that occurs over a very wide range of incidence angles (Brisco, Short, Sanden, Landry, and Raymond 2009). From a near range angle to about 20 deg (Carver et al. 1987), a portion of specularly backscattered energy is potentially returned to the sensor to make the image tone brighter. The water surface state is subject to weather conditions and

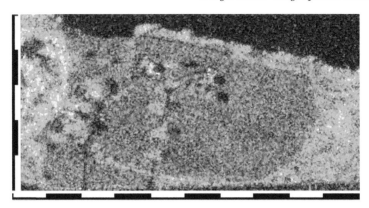

FIGURE 5.52
In a low-intensity development of Fredericton, a few homes are surrounded by trees. Both these objects seem to have comparable shadow shape and length. However, two features are different. First, the bright tone on the side of houses is much brighter and more sharply defined [(5-left,4) & (6-right,4)] than that of the trees [(5,1-high) to (5,3), & (3,4-high)]. Second, the tree tops' bright tone and paired radar shadow have near identical rounded shapes, which is not the case for houses. Shorter trees are better identified by a consistent bright tone and intermediate texture if they are grouped to create hedges or cover large area (10,1-3). On the image's north and west portions, the very bright tone of trees, bushes, and natural herbaceous indicate flooding state of a wetland (1,4-5) and a section of the Saint John River shore (5-7,5). Image: CSK 936 HH 41 deg RA Fredericton W 500 m LL 45.966451 -66.699978 deg.

is expected to affect its appearance. For instance, brighter tones, a coarser texture and patterns emerge where ice, or wind-driven, ripples or waves are formed. Once it is sited based on the tone and texture, the interpretation of a water body considers shape and dimensions as well as terrain components such as tributaries and surrounding land cover types. These aspects potentially help identify the usage of water body for a resource. The cases presented are canals and irrigation ponds, as human constructed water bodies, and naturally occurring lakes and rivers. This section presents four examples of calm surface water bodies, different by their shape. The fifth part shows images of water surfaces altered by ice and turbulence. Engineered canals and ponds have contrived shapes while rivers and lakes are naturally shaped by the terrain onto which they have developed and rest.

5.8.1   Canal

Humans design canals for purposes such as navigation and water resource management. These large infrastructures stand out in the landscape by the

FIGURE 5.53
The Pilastreti Canal [(1,3) to (4,5)] is about half the width of Napoleonica
Canal (8,1-4), yet both display similar features. The dark tone corridors have
constant and soft shape across the lands, differing from the sinuous shape of
Panaro River [(4,1) to (5,3)]. The waterways' shore segments perpendicularly
exposed to the radar beam are highlighted by a bright tone. At each canal
outlet, weirs that serve as flow regulators, hydroelectric dams, bridges, and
lock gates [(3-4,5) & (8,4-low)] stand out against the dark water background
by the very bright tone short lines, or points. Throughout the region, the land
cover includes herbaceous cultivated fields that display multiple tones, fine
texture, and patchwork pattern. Intermediate tone natural and planted forests
display a coarse (6,4) and intermediate (4,3) texture, respectively. Image: CSK
806 HH 27 deg RD Ferrara W 4.7 km LL 44.924025 11.426295 deg.

dark tone and fine texture of the water with which they are filled, and the
several kilometer straight path. The constructions are in segments and small
radius curves allow for changes of orientation and width. The Pilastresi and
Napoleonico canals (Figure 5.53) were built over 200 years ago to drain waters
from the Emilia Plain, Italy, and make the grounds appropriate for agricultural
industry to develop (Walker 1967). Along with Panaro River the canals are a
tributary to the Po River.

5.8.2   Irrigation Pond

Irrigation ponds are small, square- and rectangular-shaped water basins. They
are distributed throughout the landscape according to the agricultural effort

FIGURE 5.54
On the Lake Trasimeno watershed, Italy, an agricultural industry dominated land cover includes several irrigation ponds of up to about 1 ha in size. They are noticeable by the dark tone and very fine texture (4,4-low), which overall creates a pitted pattern on a patchwork of cultivated fields [(1,1-5) & (5,4-5)], and mature forest lands [(2,3) & (7,3)]. Their straight edges and location, coordinated with farmed fields, are in contrast with that of a stream (8-9,1) and connected lake (10,1), which are tucked in a forest. However, sections of the lake's shore appear to have been altered, indicating that it is likely also used as a reservoir. Image: CSK 908 HH 31 deg RA Cortona W 3.5 km LL 43.168070 11.985397 deg.

and its needs for water. A connection to natural streams may be noticeable, but irrigation ponds often appear isolated from drainage networks because they are refilled from ground and precipitation water. Fences and quality control equipment may be in proximity to irrigation ponds, which would cause bright tones from diffuse or corner backscattering of the radar beam (Figure 5.54).

5.8.3   Lake

Lakes exhibit irregular shapes and a wide variety of sizes. Similar characteristics emerge which identify them to a particular geographical unit. For example lakes in an alluvial plain may display smoothly curved boundaries, while those carved in surficial deposit covered bedrock may show sharp and intricate shapes consistent with their setting in depressions from glacial erosion, joints, fault escarpments, or other geomorphological features (Figure 5.55).

FIGURE 5.55
In the Gatineau Hills, lakes (3-4,4) carved in a thin colluvium blanketing the Boreal Shield bedrock have sharp angular outlines. The dark and intermediate tones, and fine texture, are distinct from the surrounding forest that is of intermediate to bright tone and intermediate texture. The brightest tones, along linear shapes, highlight abrupt slopes and hill sides exposed to the radar incident beam [(8,5) & (9,1-2)]. Some of them flank lake shores [(7-8,4) & (3,3)]. The forest hilly landscape is punctuated by roads and houses, illustrating open space developments [(1,1) & (9-10,4-5)]. Image: CSK 909 HH 18 deg RA Ottawa W 4.0 km LL 45.591903 -75.671266 deg.

5.8.4 River

Rivers are linear features of variable widths and shapes, but retaining a very high length-to-width ratio (Figure 5.56). The identification of a stream, if open water is not discernible, may be through association with the valley shape and the layout in the landscape (Figure 5.38). Built infrastructure such as bridges, dams, retaining walls, marinas and industrial wharves are often associated with rivers. Comparatively, roads and streets are also linearly shaped, narrow, and intermediate to dark tone, but unlike rivers, they have steady widths, form angular networks, and give access to building developments.

5.8.5 Water Surface State

The water surface state affects the radar backscattering and consequently the tone, image texture and patterns. Processes altering the surface of water involve ice formation and turbulence. As specular backscattering gives way to diffuse and, in some cases, corner backscattering, these features are identified by bright tones, coarser texture and patterns informative of factors affecting

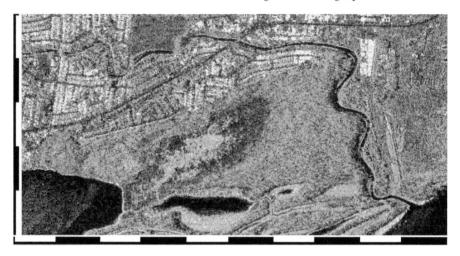

FIGURE 5.56

Rivière Blanche, Gatineau, is a low energy flow and sinuous shape meandering river. It is about 40 m wide at the confluence with Ottawa River (10,1). Most of the path represented on the image, and ponds to the south [(1-2,1) & (5,1)], have very-dark to dark tones contrasted with the nearby land cover types. The water surface marks the natural land and cityscape along almost the entire river length represented on the image. Upstream (2,5), where it is about 15 m wide, the path is sporadic due to an overhanging tree canopy. The surrounding open lands include a dense forest (7-8,2-3) and marsh land (4-5,2-3). They both display an intermediate image texture, but different tones. For the forest, it is intermediate, and for the marsh, dark and bright tones represent bare soils and hydrophyte vegetation, respectively. Medium (3-left,5) and high (2,4) intensity developments are dedicated to residential dwelling and, in a confined lot, a scrap car yard (9,4). Image: CSK 909 HH 18 deg RA Ottawa W 3.5 km LL 45.496254 -75.563064 deg.

the water surface. Three examples are proposed in this section. They include iced coastal and river water surfaces, and surface turbulence.

A dark tone is attributed to veneer ice that, like open water, helps specular backscattering. Rough surfaces, of millimeter-scale height as defined by the Rayleigh Criteria, are represented by diffuse backscattering, which increases the image tone (Ramjan, Geldsetzer, Scharien, and Yackel 2018). The tone alone may be misleading due to factors such as wind-altered open water (Shokr, Ramsay, and Falkingham 1996). In this case it may be confused with snow-covered ice and landmasses. The texture and pattern are determining keys to the interpretation of ice types. In an altered state, fragmented ice manifests bright tones, due to diffuse and double backscattering. Deformed block shapes and mottled patterns result from the ice pack being a dynamic natural environment. Ice melts and refreezes on the surface, and is subject

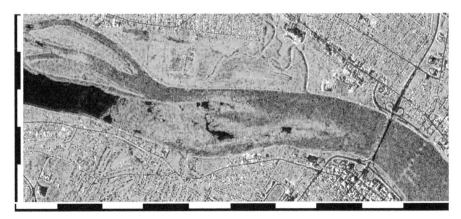

FIGURE 5.57
The image of Saint John River at Fredericton on March 26, 2009 depicts a portion of the ice jam that caused flooding upstream a few weeks earlier (City of Fredericton 2008). The very dark tone open water (1-2,3-high), bright tone pitted broken ice (4-5,3), and dark tone smoother ice (8-right,3) surface indicate different flows within the river. Longitudinal ice ridges have formed in the middle, with brighter tones [(4,3-high) & (5-right,2)] marking the transition between broken-up flows and smooth ice. Other bright tone linear features outline the river shores. Those which are perpendicularly oriented to the radar beam look direction are particularly well contrasted [(4-left,4) & (10-left,3)]. To the east, bright dots and streaks depict dragged broken ice on the downstream side of relic bridge pillars (10-left,2). Image: CSK 953 VV 41 deg RA Fredericton W 5.5 km LL 45.970171 -66.669036 deg.

to pressures by free water underneath, varying currents, confluence of other rivers, and curbing shores. Breakage of ice surfaces creates uneven striped or patchy patterns. Cracks and ponds expose open water, and like refrozen water filled fissures they are represented by dark tones, while alternating ice floes produce intermediate and bright tones. In a river environment (Figure 5.57), compared to an open bay or large lake (Figure 5.58), strong currents may cause large fissures, pressure ridges, and ice stacking. Displays of contrasted tones due to open water dark tones next to rough surface and raised ice blocks make a pattern uneven.

Surface turbulence from wind exposure brightens the tone and fine to intermediate textures combines with a streaked or rippled pattern. These may appear because the river surface current and wind stream are interacting linear forces that, in turn, affect an angularly incident radar beam to produce backscattering represented on the image. Should shoreline terrain topography shelter the water from any wind exposure, the resulting specular backscattering on calm water creates a dark tone. A sustained wind greater than 2.8 m/s modifies the water surface state enough that is can be detected on

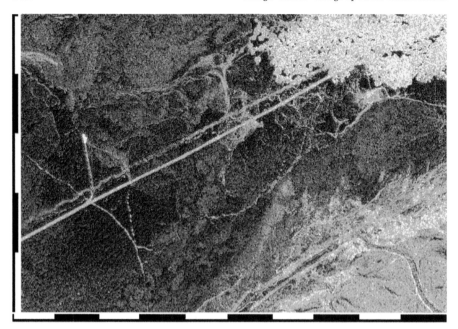

FIGURE 5.58
Northeast of Happy Valley-Goose Bay and just east of Gosling Lake, Lake
Melville presents large expanses of dark (1-2,1) and intermediate (9,3) tones
for smooth and rough ice surface textures, respectively. These characteristics are consistent with a HH-polarization representation of the ice cover
around Cornwallis Island, Nunavut, Canada (Geldsetzer and Yackel 2009).
The darker tones display fine to intermediate image textures, while the intermediate tone area exhibits a dark veined pattern (4,1). Broken ice flowing
out of the Churchill River (7-10,5) has very bright tone and coarse texture.
Human activity is indicated by a shipping lane (Atlas of Canada 2019). Broken
or rough ice filled ship wakes have a bright tone and are very well contrasted
all along their straight line crossing of less altered surfaces [(1,2) to (7,4)].
The tracks have a bright tone due to a rougher surface texture. However,
while flooded, ice may display a faded contrast due to a mix with specular
backscattering on veneer refrozen or open water. Image: CSK 208 HH 39 deg
RD Labrador W 4 km LL 53.374473 -60.226486 deg.

synthetic aperture radar images (Brisco, Short, Sanden, Landry, and Raymond 2009). Other factors that cause surface turbulence are wave-building
wind fetch, sudden changes in the underwater terrain slope, and irregular flow
regime in shallowing and narrowing sections of a river or at tributaries. A portion of Ottawa River is shown on three images acquired at one week intervals
during August 2014, on days of low (Figure 5.59), moderate (Figure 5.60) and
fair wind speeds (Figure 5.61).

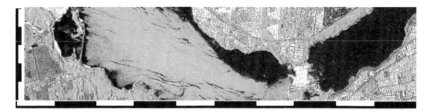

FIGURE 5.59
A west-northwest 1.4 m/s wind exposes the river section represented by a bright tone (4,3) and bordered by a streak pattern consistent with the shorelines and longitudinal river axis [(4,2) & (4-left,5)]. The wind seems to have less influence on the areas of Shirleys Bay (2-left,4), shore side waters near a large land point (7-left,2-3), and downstream from Deschênes Rapids (9,3). Image: CSK 032 HH 18 deg RA Ottawa W 13 km LL 45.376624 -75.837733 deg.

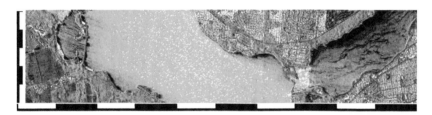

FIGURE 5.60
A westerly wind of 2.5 m/s meets with few obstacles, as indicated by a bright tone and fine texture spread over the river, except on a north shore segment (9,4) that is sheltered by Deschênes'point. Image: CSK 036 HH 18 deg RA Ottawa W 13 km LL 45.376624 -75.837733 deg.

FIGURE 5.61
The south-southwest 3.9 m/s wind is aligned with the river's downstream long axis, where patches of dark and bright tones stretch in the narrowest section [(9,3) to (10,5)]. Upstream, over half the river width, streaks show the shore interferes on the wind flow (3-4,3). Shirleys Bay (2-left,4) appears sheltered from the wind, as well as the south shore hugging waters [(3-left,2), (4-right,1) & (8-right,1)]. On all three images (Figures 5.59 to 5.61), the brightest tones and intermediate texture denote turbulent water concentrated at Deschênes Rapids, site of a decommissioned hydroelectric power station. Image: CSK 034 HH 18 deg RA Ottawa W 13 km LL 45.376624 -75.837733 deg.

5.9   Wetland

Wetlands are organic-material-dominated grounds that host grass, fen, and shrub communities, with soil sometimes developed enough to support tree stands. Water is spatially and temporally omnipresent in the form of ponds, that saturate the ground, or by forming a gradual interface with the land. The site of wetlands in depressions, level terrain, or in proximity to rivers or ponds makes them prone to flooding. This section proposes three examples of wetland environments; that of a wet meadow (Figure 5.62), string bog (Figure 5.63), and swamp (Figure 5.64).

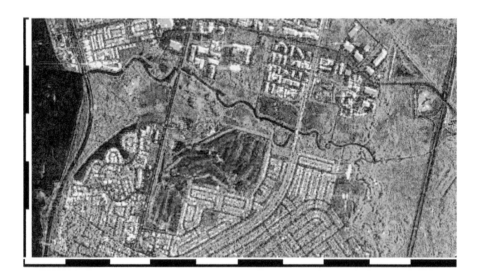

FIGURE 5.62

As a protected and precarious environment, wetlands are lightly used open spaces, as it is the case for Rivière Saint Jacques wetland (Comité Ville-Marie 2017). Downstream and near Saint Jacques River, the most water-saturated part of this wet meadow has a section of tall grass vegetation dominated by cattails and reeds. It is represented on the image by an intermediate tone and intermediate texture (2-right,3-4). The tone is brighter for inundated sections (3-right,4) & (10-left,4-low). Upstream, trees and shrubs tolerant to occasional flooding occur in larger proportions; these create a slightly brighter tone, coarser texture, and some radar shadow [(4-right,4) & (9,2)]. Trails and canals contribute to coarsening the image texture and form small dark lines [(3-left,4) & (8-left,3)]. Rivière Saint Jacques wetland is found in the middle of a high-intensity residential area [(2,1), (5-7,1), & (2,5)], outdoor recreational lands [(5,2-3) & (8,2)], and service industries (5-8,5). Tributary to the Canal de la Rive Sud (1,4), the wetland is traversed by three four-lane-divided highways [(2,3), (4,3), & (10,3)] and a railroad (7,3). Image: RS2 476 HH 35 deg RD Montréal W 4.1 km LL 45.424980 -73.474621 deg.

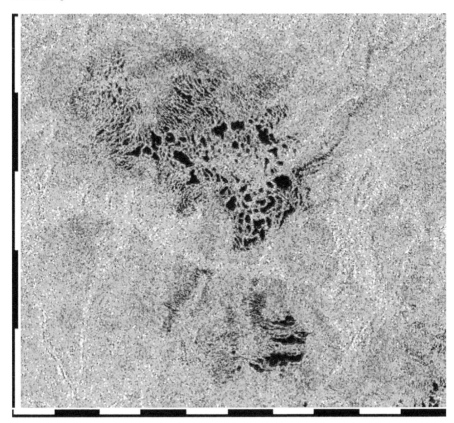

FIGURE 5.63

String bogs and ribbed fens on the coastal plains of Lake Melville (Forsyth, Innes, Deering, and Moores 2003) feature an irregular spotted pattern made by alternating ponds and fen-covered moss filaments, or developed soil (5-6,4). In the absence of large pools, a mainly fen-covered surface displays dark to intermediate tones and a coarse texture [(3,4), (4-right,2), & (5-right,1)] that gradually becomes finer toward the wetland edge where less ponding occurs [(2,4) & (4-left,2)]. Also of coarse texture, but overall much brighter in tone, is the dense coniferous forest which becomes more important at the wetland periphery [(3,3), (8-right,2-3), & (8,5)]. Trees also border Gosling Brook headwater streams (2,1-2). Image: CSK 925 HH 32 deg RA Labrador W 4.2 km LL 53.491041 -60.189416 deg.

5.10   Summary

The interpretation of large-scale geographical ensembles constitutes the essence of this chapter, with sixty-four images grouped in nine categories

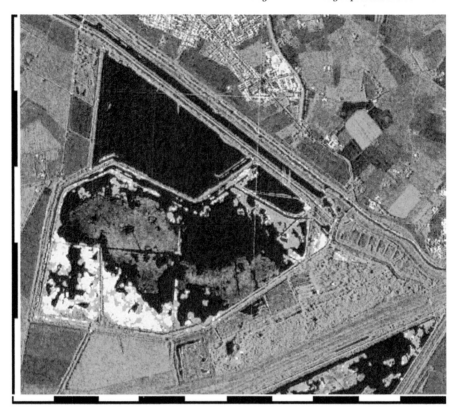

FIGURE 5.64

The wetland of Parco Regionale del Delta Del Po Park at Argenta is a protected conservation area and a series of reservoirs to collect water overflowing from surrounding rivers (Ramsar 1998). Open water (4,4) and three vegetation environments are distributed in the different basins. Floating leaf plants such as nuphar emerge at the surface of the middle basin deep water area [(3,3) to (6,2)]. Reed beds and riparian woods are in shallower waters (2-4,2) and on the middle basin banks, while a dense hygrophilous forest and shrub understory grow on a floodplain (8,2-low), that is spotted with a few ponds (6,1). Emerged plants show similar fine to intermediate textures that differentiate them from the very fine texture of water, and from the coarser texture and mottled pattern of the forest. Further, the upright shape of reeds in water is favorable to diffuse and corner backscattering (Ormsby, Blanchard, and Blanchard 1985), which creates a very bright tone, while specular backscattering is facilitated by the surface of, sometimes partly flooded, wide floating leaves. The surrounding lands are developed for agriculture [(1,4-5) & (9,4-5)] and the high-intensity development of Argenta is located just north of the park (5-6,1). A dense canal network connects the wetland interior and boundaries for draining waters. Two water-filled canals border the marsh north side [(2,5) to (10,2)]. Image: CSK 006 HH 39 deg LD Ferrara W 4.7 km LL 44.593843 11.831891 deg.

involving barren lands, developments, producers and services industries, landforms, recreational land use, vegetation, water bodies, and wetlands. Rubrics introduces the topic of interest, some of the physical attributes that one expected to be seen, but more importantly, that is likely to affect the radar beam backscattering, and a description encompassing relevant interpretation criteria.

A synthetic aperture radar image subset represents each ensemble, and location coordinate-assisted narrations point to visual information and LULC interpretation. The image geographical extents are relatively limited, up to 20 km. This, with like-horizontal polarization images for the majority of examples, helps keep a consistent representation throughout the chapter. Preprocessing was kept to a minimum in order to maintain the images' essential details. Coordinates locate the examples within each image and the figure caption's closing statement provides ancillary information about the image acquisition specifications, place name, area width, and geographic location.

Large-scale representations give opportunities to exploit the interpretation keys of tone, texture, pattern, shadow, shape, and dimensions, as a multivariate information set. Several visual examples help to familiarize with 'reading' synthetic aperture radar images. A systematic strategy for proposing interpretation solutions can be applied with other images acquired from active radar sensors.

Compared to regional-scale applications presented in Chapter Four, the texture, shape, and dimensions are particularly useful keys to the interpretation of objects and portions of geographical ensembles which size are of the same order as the image spatial resolution. A variety of texture can be observed on continuum before patterns become apparent, and relatively small objects and areas can be outlined by their contrasted tones that emerge from the high spatial resolution image products.

6

Image Acquisition Specifications into Effect

This section presents application and image examples investigating different acquisition specifications. Synopses consider the incidence angle, look direction, polarization and spatial resolution. Highlights are provided for guidance on interpreting the land cover and use from different synthetic aperture radar image perspectives, while reporting more specifically on the backscattering relative responses.

6.1   Incidence Angle

As land cover broad categories, natural and urban environments are effectively interpreted from particular ranges of incidence angles. Backscattering value analyses reveal that near range acquired images, from incidence angles near 30 deg, are more useful to discriminate between vegetation land cover categories such as forest, wetland and planted herbaceous (Bernier, Ghedira, Gauthier, Magagi, Filion, Sève, Ouarda, Villeneuve, and Buteau 2003). Far range images, with incidence angle of about 45 deg, help differentiate woody bogs from surrounding forests, mainly because the low grazing radar beam more easily interacts with their changing structural characteristics. Open water is well contrasted to wet and non-flooded lands from 50 deg incidence angle HH polarization images, during ideal low surface wind conditions (Töyrä, Pietroniro, and Martz 2001). However, water surfaces modified by moderate wind velocities are differentiated from land with lower angle and cross-polarization options. In urban areas, incidence angle effects on the backscattering intensity are observed as various land cover components act on allowing one of corner, specular, or diffuse radar beam returns to dominate. A homogeneous pattern created by the combination and sequence of tone and texture is formed for urban land use types. A small incidence angle facilitates the detection of street and road infrastructure in a high-intensity building fabric, especially where the buildings are of moderate height (Chen, Zhang, Guindon, Esch, Roth, and Shang 2013). A large incidence angle helps to counteract image layover at the location of tall buildings, but meets conditions for foreshortening-related saturation to occur on building sides facing the incident beam, and radar shadow to be formed on the opposite side.

The interpretation of mountainous, hilly, and gently rolling topographies' structural components must consider knowledge of the terrain slope that would likely be encountered (Singhroy and Saint-Jean 1999). Acquired from far ranges, of 40 to 60 deg incidence angles, images are favorable for the interpretation of mountainous landforms because radar beam exposed slopes and incidence angles are similar. Moderate topography geomorphological features, where the slopes range between 25 and 35 deg, are best enhanced by 20 to 30 deg incidence angle images. These prevail for enhancing linear and glacial features such as those of the Boreal Shield Ecozone, alluvial, morainic and delta deposition environments, and small river valleys. Near range images, of say 18 to 25 deg incidence angle, work to visualize level and moderate terrain shapes. To have an incidence angle similar to the slope facing it would have the land cover representation affected by foreshortening, which helps locate small inclines. Opportunities for land cover interpretation, in addition to terrain shape, are identified by using images acquired from incidence angles that are large enough to allow for surface roughness to take effect. Sometimes, on abrupt hillsides, this is only possible by using a small incidence angle to view slope sides that face away from the incident radar beam. Essentially, land cover on flat terrain is potentially interpreted from a wide range of incidence angles. For topographies presenting more abrupt slopes, limitation to very large incidence angles is the choice of images that are useful for land cover interpretation. These create the conditions for radar shadow to occur. Images from two different look directions or orbital paths potentially constitute complementary data for the interpretation of land cover.

Small and large angle of incidence used in a numerical ratio show strong correlation to variation of soil material size values from 2.5 to 5.5 cm (Sahebi, Angles, and Bonn 2002) which is within the X-band (Table 2.1) and C-band (Table 2.2) rough-surface texture range. Applied in a visual interpretation context, images acquired from extreme incidence angles, say of 20 and 45 deg, would be expected to represent the different materials in a discernible range of tones. However on a single image that has a incidence angle range of a few degrees, tonal changes would set apart general material size groups, as defined by the three Rayleigh Criterion categories.

Herbaceous planted vegetation displays a variety of tones built in patchwork patterns. Factors such as the crop type, growth stage, and water content affect the tonal value, and the contrasts are found to be more important on images acquired at incidence angles of greater than 30 deg (Gherboudj, Magagi, Berg, and Toth 2011). Yet, information about planted row structure and background soil must consider moderate incidence angles or image acquisition at a non-dominant canopy coverage period (Mathieu, Sbih, Viau, Anctil, Parent, and Boisvert 2003). Research conducted in Manitoba, Canada (McNairn, Ellis, Van Der Sanden, Hirose, and Brown 2002) demonstrates the ability to classify the grain crop types, assess the growth by four stages: emergence, vegetative, headed and senescence, and differentiate canola crops' flower and pod set stages. Three indicators of crop vigor were retained for detailed investigation: the leaf area index, wet biomass, and plant height. A multitemporal dataset included five incidence angle beam modes spanning from the near to far range, with 20 to 48 deg. The optimal crop type classification is based on images acquired during the period of plant seed development and, using fine mode with a large incident angle, greater than 35 deg. A multitemporal image set is required for a detailed classification of the crop types and the characterization of their growth stage. A dataset that encompasses the entire growing season conveys backscattering variation at many development phases of wheat, barley and oats grain crops, and also the differentiation of these from corn and sunflower leaf crops. The relationship of backscattering to the crop vigor variables is strongest for wheat and potato fields. The backscattering values of canola crops on a mid-season 40 deg incidence angle image are higher than those of grain crops at any growing stage. From the flowering time to pods being developed, the backscattering values of canola crops increase. Meanwhile, the grain crop backscattering decreases as the growth progresses from emergence to headed, then increases at the senescence stage. The crop stage relationship to the backscattering is believed to be affected by the changing exposure of soil between rows of vegetation. Once crop canopies dominate in term of coverage, the indicators of crop vigor take precedence. In particular, the increase of leaf area index and height of wheat, potatoes, and canola crops causes higher backscattering to be recorded. Exceptions noted for wheat crops on the mid-season images are explained by the structural change of the plants and interaction of the background soil. The radar beam sensitivity to the physical configuration of crops is captured while considering different incidence-angle images, multi-temporal series, and contrasted spatial resolutions, for example, in the context of an object-based classification (Jiao, McNairn, Shang, Pattey, Liu, and Champagne 2011).

6.2   Look Direction

The synthetic aperture radar beam look direction setting influences the appearance of land cover and objects that have a vertical dimension and form alignments in the landscape. Despite the spatial distortions it introduces this image specification benefits the interpretation process. Image tones are generally brighter on the sides exposed to the radar beam and darker where radar shadow is formed on the far range object facets. This configuration is controlled by selecting a right or left look direction from an ascending or descending orbital path. The effect on tone, texture and dimensions is emphasized by using large incidence angles. Over a large region, the look direction gives different perspectives on any given terrain shape. Where it is relatively flat, little change is perceptible in the shapes and relative position of objects about each other. In sloping terrain, changing local incidence and azimuthal angles create distorted dimensions over the landscape. A local incidence angle is smaller on radar beam exposed terrain slopes and larger on the opposite side, until shadow occurs. The look direction affects the perspective, but as long as the slope angle of the terrain exposed to the beam remains well below the beam incidence angle, the local incidence angle keeps relatively low, making the image suitable for applying visual interpretation keys and extracting land cover information. Opposite look directions from descending (Figure 6.1) and ascending (Figure 6.2) orbits appear rotated by a few degrees from each

FIGURE 6.1
Image: CSK 212 HH 31 deg RD Cortona W 3.4 km LL 43.259515 11.980383 deg.

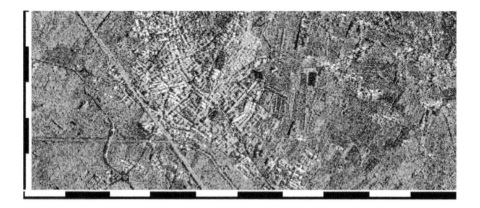

FIGURE 6.2
Image: CSK 908 HH 31 deg RA Cortona W 3.4 km LL 43.262109 11.983163 deg.

other. They produce complementary coverage from which to interpret the land cover. Images acquired from a right looking direction, from descending and ascending orbits, have their near range toward the East and West, respectively. The local incidence angle on West oriented slopes is smaller on the right ascending image than on the right descending image. In instances of a favorable azimuthal orientation, sharper contrasts occur for the high density development on the right descending image, while very bright saturated tone and coarse texture blurs some of the buildings aligned perpendicularly to the right ascending radar beam. The radar beam look direction allows enhancement of different urban landscapes' portions because of the built infrastructure various orientations and densities, and changing building dimensions. Level-ground vegetation canopies, calm water and large asphalted surfaces are little affected by the change in look direction.

From opposite look directions, large units of valley agriculture (1,1-5), urban developments (4,4), and permanent crops (7-10,2-5) are differentiated. Objects in the valley (1-3,1-5) have similar tones, shapes and distribution on right look images acquired from descending (Figure 6.1) and ascending (Figure 6.2) orbits. The azimuthal orientation change is shown by a straight railroad crossing the image West portion. High vertical objects are identified by the shadow on their far range side, for example, a large bridge [RD: (3-left,2) & RA: (2,3-low)] and a multistory building [RD: (3-right,2-high) & RA: (3,3-low)]. West-facing hillside sparsely-treed lots are darker on the right look descending orbit image (9-10,2-4). At this location a larger local incidence angle minimizes foreshortening and makes it easier to identify low-intensity developments. The changing topography has the local incidence angle change and spatially affect radar shadow occurrences and foreshortening or corner backscattering on building walls. This helps explain tone and texture difference in the permanent crop appearance [RD: (8-9,4) & RA: (7-8,3-4)] and the build environment [RD: (5,4) & RA: (5,3)]. Smooth texture surfaces such as asphalt [RD: (6,4) & RA: (6-left,3)], ponds [RD: (8,3-low) & RA: (7-right,1)] and lawns [RD: (6-right,4) & RA: (7-left,2)] display a very dark tone on both the images. Across hills, a narrow local incidence angle enhances a road from [RA: (6,4) to (8,4-2)].

Comparing opposite look images helps with the interpretation of single or linear objects such as tree rows and edges, row crops, buildings walls, gabled roofs, hill sides, bridges, and depressions (Figures 6.3 and 6.4).

Right looks from descending (Figure 6.3) and ascending (Figure 6.4) orbit images reveal a succession of terraces inside an excavation (2,2-3), and v-shape valley slopes [(4,1-5), RD: (8-9,1-5) & RA: (8,1-5)]. A bright line marks the slope that has a value about that of the beam angle, or a local incidence angle near zero. A radar shadow is on the side of terrain slopes turned away from the onlooking beam. A bright-dark sequence is observed from the image far to near range for negative slopes such as canal and valley sides, and excavations walls. Objects above ground such as tree lines, buildings, and raised embankments

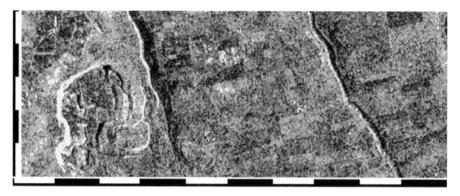

FIGURE 6.3
Image: CSK 840 HH 31 deg RD Taranto W 3.2 km LL 40.598365 17.183103 deg.

FIGURE 6.4
Image: CSK 055 HH 29 deg RA Taranto W 3.2 km LL 40.601644 17.182795 deg.

are enhanced by a bright-dark sequence in the opposite direction, from near to far range, for example, a u-shape sediment pile on the excavation East side [RD: (4-left,1-2) & RA: (3,1)]. Where the terrain is relatively flat, alike tones, textures and patterns are represented on the right look descending and ascending orbit images, including the bare excavated grounds in enclosed curved shapes formed by different level terraces, agricultural patchwork of fields, some marked by dotted and striped patterns, and coarsely textured uneven vegetation overgrown river valley segments.

6.3 Polarization

Tone and texture image interpretation keys are beneficial while considering the like-polarization, HH or VV, and crossed, with HV or VH, as two main options. The former conveys surface backscattering information and the latter differentiates volumes or structures that cause the radar beam to change orientation (Maghsoudi, Collins, and Leckie 2012). Like- and cross-polarization image pairs represent generally compatible information, at least visually. The HH and VV images for the areas depicted in this section are 50 % correlated, while the HV and VH numerical values do fluctuate very strongly about each other as they are 90 % correlated. Therefore, the land cover type differentiation with regard to this synthetic aperture radar image acquisition parameter is based on like- and cross-polarization option summaries, prepared by the mathematical products of the HH and VV options, and HV and VH, respectively.

Acquired with Radarsat-2 Fine Quad for Fredericton (Figures 6.5 and 6.6) and Standard Quad for Labrador (Figure 6.7 and 6.8), the images are regional scale examples that like- and cross-polarization provide complementary land cover interpretation solutions.

The like-polarization image (Figure 6.5) conveys sharp contrasted bright tones related to buildings of the Mactaquac Dam site (2,2-high) and high-intensity development on both sides of Saint John River (8-9,1-2), abrupt hillsides [(4,5), (6,2), & (8,4)], and plowed crop fields (4-5,3-high). The cross-polarization image (Figure 6.6) also highlights some of the hillsides and large buildings, but in addition, some unevenness of the forest canopy that in places [(3-4,1) & (10-right,3)] appears as a slightly coarser texture on the cross-polarization image. This indicates the presence of multi-faceted leaves, branches and forest canopy openings causes the radar beam polarization emitted orientation to change. With the low incidence angle this image is acquired with, the cross-polarization brings out better contrasts between dark-tone roads and intermediate-tone transmission lines of all directions on the bright tone and forest land background.

Differences between like- (Figure 6.7) and cross-polarization (Figure 6.8) images are noticeable for water surfaces, wetlands and open forest. As with the previous example, but from a right look and descending orbit, the radar beam facing slope breaks are more contrasted on the like-polarization image. At the time of image acquisition, the wind is reported to be 5.6 m/s northeasterly, which is a higher velocity than in previous examples (Figure 5.61). On the like-polarization image, Lake Melville (8-9,1-3), Gosling Lake (5-6,2), and ponds scattered throughout the landscape that are large enough for a wind fetch to cause turbulence are represented by an intermediate tone. The smaller ponds display a dark tone.

On the cross-polarization image, bogs form large dark tone patches with some darker specks for the ponds, which are very well contrasted with the

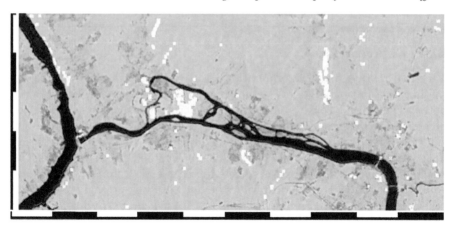

FIGURE 6.5
Image: RS2 086 Like-polarization 21 deg RA Fredericton W 25 km LL 45.981003 -66.764274 deg.

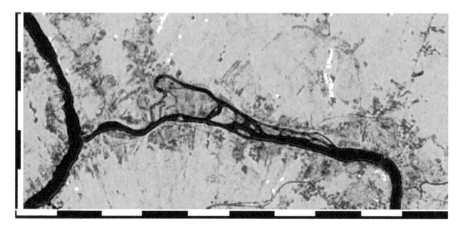

FIGURE 6.6
Image: RS2 086 Cross-polarization 21 deg RA Fredericton W 25 km LL 45.981003 -66.764274 deg.

surrounding dense and fire-scar open forests. Conversely, only the treed bog portions stand out on the like-polarization image by their bright tone. The fens and moss areas are intermediate tone, less contrasted to surrounding forests. Wetland patches are distributed throughout the Lake Melville Plain [(7,3) to (8,5)]. Where the tone is dark on both the like- and cross-polarization images, for example on the Gosling River North shore (6-7,2) and uplands (5,4-5), the

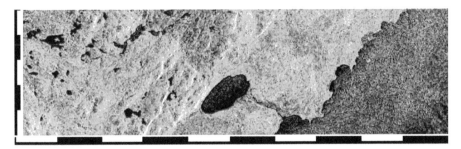

FIGURE 6.7
Image: RS2 475 Like-polarization 24 deg RD Labrador W 27 km LL 53.428732 -60.365974 deg.

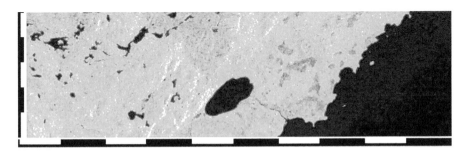

FIGURE 6.8
Image: RS2 475 Cross-polarization 24 deg RD Labrador W 27 km LL 53.428732 -60.365974 deg.

land is mainly grass covered or barren grounds. On both the images, forest lands display a coarse texture and intermediate to bright tones.

6.4 Spatial Resolution

The spatial resolution is a system specification that similarly affects the ability to interpret LULC types from synthetic aperture radar and multispectral images. Small geographic scale land cover is assessed with both the Radarsat-2 Fine and Standard mode images that are 10 m (Figure 6.9) and 20 m (Figure 6.10) nominal spatial resolution. Forest lands have a bright tone, wetlands and barren lands are dark, and developed land features dark tone streets and small bright points at the location of buildings. Interpretation keys of tone, texture, pattern, and shapes are rendered with more contrast and sharper

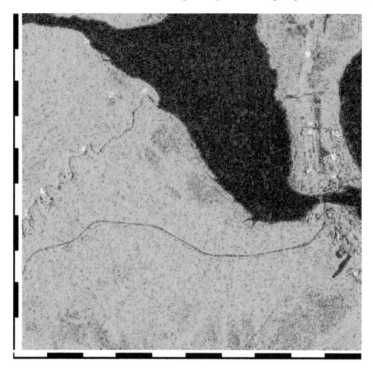

FIGURE 6.9
10 m Spatial Resolution Image: RS2 474 Four-polarization PCA 37 deg RD
Labrador W 5.6 km LL 53.529026 -60.177717 deg.

boundary definition on the higher spatial resolution image. Forested, barren,
and wet lands, which form areas, display coarser textures on the Standard
mode image, but bodies of water have, as expected, a fine texture on both
the images. Continuous linear shapes are just about perceptible on the low
spatial resolution image. A sinuous river path and regional road are sporadically mingled with the forest. This confusion is more noticeable where their
orientation departs from the radar beam look direction (1-3,3-4). A relatively
high incidence angle, of 37 deg, creates an additional difficulty for the radar
beam in reaching small asphalted roads or open water surfaces surrounded by
trees and hillsides.

Other landscape features are enhanced on both the image resolutions.
Radar beam foreshortening produces contrasts that help identify some of the
river meander banks [(2,5) & (3,6)] and a terraced escarpment (9,9-10). Transmission line corridors modestly impress on the forest and wetlands on the high
spatial resolution image only [(7,1) to (10,2) & along the road's south-side].
The low-intensity developments of Sheshatshiu (9-10,3-4) and North West
River (9-10,6-7) are interpreted mainly due to the contrasted set of dark tone

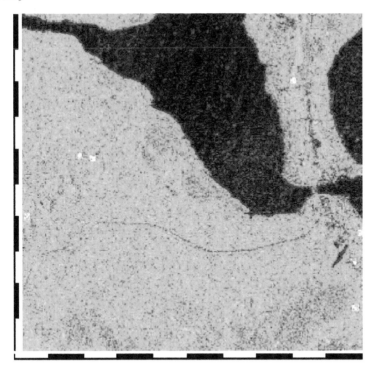

FIGURE 6.10
20 m Spatial Resolution Image: RS2 475 Four-polarization PCA 24 deg RD
Labrador W 5.6 km LL 53.529026 -60.177717 deg.

lines across bright and intermediate tones areas. A higher spatial resolution
facilitates the interpretation of smaller components. A few very narrow streets
are noticed, some of the buildings are brighter in tone, and a larger variety
of tones in the built-up fabric suggest parking areas, open spaces, and small
ponds.

6.5 Summary

This chapter offers selected research outcomes and contributes visual examples
based on divergent specifications for synthetic aperture radar image acquisi-
tions. Discussions and examples consider the incidence angle, look direction
and orbital path, polarization mode, and spatial resolution. Visual effects of
acquisition specifications are shown through outlined differences and similarities in the appearance of LULC types.

The observations concur, a large incidence angle and coarse spatial resolution assist the differentiation of open land and natural environment categories, i.e., forests, topography, water wetlands, which are intrinsically vast and form low spatial variability units. Conversely, near-range image positioning and fine spatial resolution options facilitate the interpretation of developed LULC and small dimension landscape features, such as buildings, transportation axes and linear waterways, and boundary environments, such as terraces, cliffs, riverine and marine coastlines.

Some of the look direction and incidence angle alternatives introduce spatial distortion and obstructions in mountainous, hilly terrain, and high-intensity-high-rise building developments. However, as part of image interpretation and for assessing shapes and patterns, these inconsistencies create information useful to the assessment of object height, topography, and surface roughness. For example, the look direction and incidence angle, in addition to the radar wavelength, influence surface and volume backscattering qualities characteristic to open water and vegetation land cover types, respectively.

7

Conclusion

This book is designed to help students, the public and practitioners who wish to use synthetic aperture radar images in their work. Familiarization with the information this image type provides is a gateway to knowledge gathering, and, subsequently, geographical or environmental problem-solving. As part of many Earth observation strategies available today, synthetic aperture radar is a reliable source of data. Exploiting electromagnetic radiation energy bands beyond the visible range is key to widening publicly available remote sensing databases. Up-to-date and historical image series can only improve the experience of interactive-learning with spatial information applications, to which access is granted through a better visual understanding of various image types.

Following the Introduction, Chapter Two presents synthetic aperture radar image characteristics that influence the appearance of objects in their visual representation. Chapter Three explains how to consider and what knowledge is conveyed by the visual keys of tone, texture, pattern, shadow, shape, and dimensions assessed specifically from synthetic aperture radar images. The book's core chapters, i.e., Fourth and Fifth, offer regional and large scale image interpretation solutions, respectively. Over 70 annotated examples feature land use and land cover types using geographical ensembles of locations in Canada and Italy. Chapter Six gives comparative examples of acquisition specification effects on objects' spatial representation.

The book construction allows for quick reference to specific topics. Image representation examples address several different land cover and land use types, some of them in various contexts. Annotations emphasize observations in an organized fashion. Interpretation keys lead to visual description applicable to different spatial scales, perspectives, and locations. The overall interpretation exercise for each object, ensemble or regional geography is repeatable. Attention to detail is important, but so is the consideration of image acquisition specifications. In the same way that multispectral image interpretation relies on knowledge of signatures and the bands being represented, synthetic aperture radar imagery must consider the acquisition parameters that bring together visual information into an image.

The examples, annotated for enhancing a particular topic, explain key features that may be used to learn about the land use or land cover identity. It is beneficial to look at different examples of a particular geographic ensemble to realize commonalities and deductions. Not only is the central point of each image important but also their surroundings, which serve as a comparative reference. Several repetitive and attentive observations of land cover and types, in different geographical contexts, help build that knowledge so important to the fluid use of synthetic aperture radar images.

Modern numerical classification methods exploit the object as a whole entity for information extraction, rather than a per pixel approach. Object-based classification involves setting a generic spatially coherent multivariable description that ultimately links it to concrete land use or cover types (Scharien, Segal, Nasonova, Nandan, Howell, and Haas 2017), traditionally referenced as an informational class. Visually understood synthetic aperture radar images at the outset help conceptualizing machine-learned object key components. This book presents synthetic aperture radar image examples of geographical ensembles that in addition to tone and texture, have distinguishing features that remain to be derived numerically in object classification processes for recreating intuitive interpretation keys such as pattern, shapes, and dimensions. Further, temporality can be envisaged as a viable component of this framework. Synthetic aperture radar imagery built in a time series, advantageously serves to interpret land use and cover. Changing soil exposure and variable tree canopy state, have been recognized as classification factors for vegetation land cover types. Images of water bodies, wetlands, and built environments also show the potential of surface dynamics time series as an information-rich variable to describe land cover types.

Image interpretation starts with seeing to describe, then proceeds into examining isolated and spatially expressed characteristics, and concludes with an improved understanding of geographies. Applying synthetic aperture radar image interpretation keys with already strong significance to information extracted from optical data will expand on currently available landscape representations. With subsequent translation to digital analysis strategies, the process becomes more effective and enables further exploitation of digitally accessible concepts. Increased spatial coverage will ensue, but also inevitably, discoveries.

Bibliography

Abbott, K., B. Leblon, G. Staples, D. Maclean, and M. Alexander 2007. Fire danger monitoring using Radarsat-1 over northern boreal forests. *International Journal of Remote Sensing 28*(6), 1317–1338.

ACPB 2017. Bois papineau, un joyaux écologique. Technical report, Laval: Association pour la Conservation du Bois Papineau.
http://www.boispapineau.ca/fr/joyau.sn#.XSSUzf57mUk/ (accessed July 2019).

Agenzia Spaziale Italiana (Eds.) 1987. *Cosmo-SkyMed SAR products handbook.* Roma: ASI.
http://www.cosmo-skymed.it/docs/ASI-CSM-ENG-RS-092-A-CSKSAR products handbook.pdf (accessed July 8, 2019).

Ahern, F., D. Leckie, and J. Drieman 1993. Seasonal changes in relative C-band backscatter of northern forest cover types. *IEEE Transactions on Geoscience and Remote Sensing 31*(3), 668–680.

Atlas of Canada 2010. *Land cover (generalized)*, Volume 6. Ottawa: Natural Resources Canada.
https://doi.org/10.4095/301036 (accessed July 8, 2019).

Atlas of Canada 2019. *Toporama.* Ottawa: Natural Resources Canada.
http://atlas.gc.ca/toporama/en/index.html (accessed July 8, 2019).

Baghdadi, N., M. Bernier, R. Gauthier, and I. Neeson 2001. Evaluation of C-band SAR data for wetlands mapping. *International Journal of Remote Sensing 22*(1), 71–88.

Ban, Y. and A. Jacob 2013. Object-based fusion of multitemporal multiangle ENVISAT ASAR and HJ-1B multispectral data for urban land-cover mapping. *IEEE Transactions on Geoscience and Remote Sensing 51*(4), 1998–2006.

Beaudoin, A., P. Bernier, P. Villemaire, L. Guindon, and X. J. Guo 2017. Tracking forest attributes across Canada between 2001 and 2011 using ak nearest neighbors mapping approach applied to MODIS imagery. *Canadian Journal of Forest Research 48*(1), 85–93.

Bernetti, I. and N. Marinelli 2010. Evaluation of landscape impacts and land use change: a Tuscan case study for CAP reform scenarios. *Aestimum 2010*(56), 1–29.

Bernier, M., H. Ghedira, Y. Gauthier, R. Magagi, R. Filion, D. D. Sève, T. B. Ouarda, J.-P. Villeneuve, and P. Buteau 2003. Détection et classification de tourbières ombrotrophes du Québec à partir d'images Radarsat-1. *Journal Canadien de Télédétection 29*(1), 88–98.

Blott, S. and K. Pye 2012. Particle size scales and classification of sediment types based on particle size distributions: Review and recommended procedures. *Sedimentology 59*(7), 2071–2096.

Bossard, M., J. Feranec, and J. Otahel 2000. *CORINE land cover technical guide Addendum 2000*. Copenhagen: European Environmental Agency. https://www.eea.europa.eu/publications/tech40add/at_download/file (accessed July 8, 2019).

Bouman, B. and D. Hoekman 1993. Multi-temporal, multi-frequency radar measurements of agricultural crops during the Agriscatt-88 campaign in The Netherlands. *International Journal of Remote Sensing 14*(8), 1595–1614.

Brisco, B., N. Short, J. v. d. Sanden, R. Landry, and D. Raymond 2009. A semi-automated tool for surface water mapping with Radarsat-1. *Canadian Journal of Remote Sensing 35*(4), 336–344.

Broster, B. 1998. Aspects of engineering geology at Fredericton, New Brunswick. *Urban Geology of Canadian Cities 42*, 401–408.

Carver, K. et al. 1987. *Earth Observing System Reports: Synthetic Aperture Radar, Instrument Panel Report*, Volume IIf. Washington: NASA.

Chen, Z., Y. Zhang, B. Guindon, T. Esch, A. Roth, and J. Shang 2013. Urban land use mapping using high resolution SAR data based on density analysis and contextual information. *Canadian Journal of Remote Sensing 38*(6), 738–749.

City of Fredericton 2008. *1973 flood vs. 2008 flood*. Fredericton: Planning & Policy Division. http://www.fredericton.ca/sites/default/files/public-safety/1973vs2008 floodmap.pdf (accessed July 8, 2019).

Comité Ville-Marie 2017. *Accès fleuve: Aménagement d'une frayère sur la plaine inondable de la rivière Saint-Jacques*. Longueuil: Comité Ville-Marie. http://www.zipvillemarie.org/ameacutenagementdunefrayegraveresurla-riviegraveresaintjacques.html (accessed July 8, 2019).

Ecological Stratification Working Group 1996. *A national ecological framework for Canada*. Ottawa: Canadian Soil Information System.

http://sis.agr.gc.ca/cansis/publications/ecostrat/cad_report.pdf (accessed July 8, 2019).

Estes, J., E. Hajic, L. Tinney, et al. 1983. Fundamentals of image analysis: Analysis of visible and thermal infrared data. *Manual of Remote Sensing 1*, 987–1124.

ESTR Secretariat 2016. Mixedwood Plains Ecozone+ evidence for key finding summary. *Canadian biodiversity: ecosystem status and trends 2010, Evidence for key findings summary report 2016*(7), 145.

Forsyth, J., L. Innes, K. Deering, and L. Moores 2003. *Forest ecosystem strategy plan for forest management District 19, Labrador/Nitassinan, 2003–2023*. Sheshatshiu: Innu Nation and St. John's: Government of Newfoundland and Labrador.
http://www.mae.gov.nl.ca/env_assessment/projects/Y2003/1062/text.pdf (accessed July 8, 2019).

Ge, P., H. Gokon, and K. Meguro 2019. Building damage assessment using intensity SAR data with different incidence angles and longtime interval. *Journal of Disaster Research 14*(3), 456–465.

Geldsetzer, T. and J. Yackel 2009. Sea ice type and open water discrimination using dual co-polarized C-band SAR. *Canadian Journal of Remote Sensing 35*(1), 73–84.

Gherboudj, I., R. Magagi, A. Berg, and B. Toth 2011. Soil moisture retrieval over agricultural fields from multi-polarized and multi-angular Radarsat-2 SAR data. *Remote Sensing of Environment 115*(1), 33–43.

Gioia, D., M. Bavusi, P. Di Leo, T. Giammatteo, and M. Schiattarella 2016. A geoarchaeological study of the Metaponto Coastal Belt, southern Italy, based on geomorphological mapping and GIS-supported classification of landforms. *Geografia Fisica e Dinamica Quaternaria 39*, 137–148.

Government of Canada 2014. *Service industries*. Ottawa: Government of Canada.
https://www.ic.gc.ca/eic/site/si-is.nsf/eng/home (accessed July 8, 2019).

Homer, C., J. Dewitz, J. Fry, M. Coan, N. Hossain, C. Larson, N. Herold, A. McKerrow, J. VanDriel, J. Wickham, et al. 2007. Completion of the 2001 national land cover database for the conterminous United States. *Photogrammetric Engineering and Remote Sensing 73*(4), 337.

Homer, C., C. Huang, L. Yang, B. Wylie, and M. Coan 2004. Development of a 2001 national land-cover database for the United States. *Photogrammetric Engineering and Remote Sensing 70*(7), 829–840.

Hydro Québec 2019. *Hydroelectric generating stations.* Montréal: Hydro Québec.
http://www.hydroquebec.com/generation/centralehydroelectrique.html (accessed July 8, 2019).

Jiao, X., H. McNairn, J. Shang, E. Pattey, J. Liu, and C. Champagne 2011. The sensitivity of Radarsat-2 polarimetric SAR data to corn and soybean leaf area index. *Canadian Journal of Remote Sensing 37*(1), 69–81.

Kanata North Business Association 2018. *Overview and mandate.* Kanata: KNBA.
http://www.kanatanorthba.com (accessed July 8, 2019).

King, R. 2015. *The industrial geography of Italy.* New York: Routledge.

Leish, J. and J. Morrall 2014. Evolution of interchange design in North America. In Transportation Association of Canada (Ed.), *Geometric design-learning from the past*, Number 10415 in Conference of the Transportation Association of Canada.
http://onf.tac-atc.ca/english/annualconference/tac2014/s30/morrall.pdf (accessed July 8, 2019).

Lewis, A. and F. Henderson 1998. Radar fundamentals: The geoscience perspective. In *Principles and Applications of Imaging Radar, Manual of Remote Sensing*, pp. 131–181. New York: John Wiley & Sons, Inc.

Lillesand, T. and R. Kiefer 1994. *Remote sensing and image interpretation.* New York: John Wiley and Sons, Inc.

Liverman, D. 1997. Quaternary geology of the Goose Bay area. Technical Report 97-1, St. John's: Department of Mines and Energy, Newfoundland and Labrador.
https://www.nr.gov.nl.ca/nr/mines/geoscience/publications/currentresearch/1997/liverman.pdf (accessed July 8, 2019).

Liverman, D. and K. Sheppard 2000. *Landforms and surficial geology of the North West River map sheet (NTS 13F9).* St. John's: Department of Mines and Energy.
https://www.nr.gov.nl.ca/nr/mines/maps/surfmap/Lab/html_lb/images/map200042.pdf (accessed July 8, 2019).

MacDonald Dettwiler Ltd 2018. *Radarsat-2 product description.* Richmond: Maxar Technologies Ltd.
https://mdacorporation.com/docs/default-source/technical-documents/geospatialservices/52-1238_rs2_product_description.pdf (accessed July 8, 2019).

Maghsoudi, Y., M. Collins, and D. Leckie 2012. Polarimetric classification of Boreal forest using nonparametric feature selection and multiple classifiers.

International Journal of Applied Earth Observation and Geoinformation 19, 139–150.

Mathieu, R., M. Sbih, A. Viau, F. Anctil, L. Parent, and J. Boisvert 2003. Relationships between Radarsat SAR data and surface moisture content of agricultural organic soils. *International Journal of Remote Sensing 24*(24), 5265–5281.

McGrew, J., A. Lembo, and C. Monroe 2014. *An introduction to statistical problem solving in geography*. Long Grove: Waveland Press, Inc.

McNairn, H., J. Ellis, J. Van Der Sanden, T. Hirose, and R. Brown 2002. Providing crop information using Radarsat-1 and satellite optical imagery. *International Journal of Remote Sensing 23*(5), 851–870.

Miranda, B., B. Sturtevant, I. Schmelzer, F. Doyon, and P. Wolter 2016. Vegetation recovery following fire and harvest disturbance in central Labrador-A landscape perspective. *Canadian Journal of Forest Research 46*(8), 1009–1018.

Murtha, P. 2000. Surficial geology and climatic effects on forest clearcut tone in Radarsat images of Northern Vancouver Island. *Canadian Journal of Remote Sensing 26*(3), 253–262.

Ormsby, J., B. Blanchard, and A. Blanchard 1985. Detection of lowland flooding using active microwave systems. *Photogrammetric Engineering and Remote Sensing 51*(4), 317–328.

Peake, W. and T. Oliver 1971. The response of terrestrial surfaces at microwave frequencies. Technical report, Ohio State University Columbus Electroscience Laboratory.

Ramjan, S., T. Geldsetzer, R. Scharien, and J. Yackel 2018. Predicting melt pond fraction on landfast snow covered first year sea ice from winter C-band SAR backscatter utilizing linear, polarimetric and texture parameters. *Remote Sensing 10*(10), 1603.

Rampton, V. 1984. *Surficial geology, New Brunswick map*. Ottawa: Geological Survey of Canada. https://doi.org/10.4095/119734 (accessed July 9, 2019).

Ramsar 1998. Valle Campotto e Bassarone - Site number 181. Technical report, Ramsar Sites Information Service. http://rsis.ramsar.org/ris/181 (accessed July 9, 2019).

Ramsar 2001. Mer Bleue conservation area - Site number 755. Technical report, Ramsar Sites Information Service. http://rsis.ramsar.org/ris/755 (accessed July 9, 2019).

Raney, R. 1998. Radar fundamentals: technical perspective. In *Principles and Applications of Imaging Radar, Manual of Remote Sensing*, pp. 9–130. New York: John Wiley & Sons, Inc.

Richard, S. 1982. *Surficial geology, Ottawa, Ontario-Québec*. Ottawa: Geological Survey of Canada.

Robert, A. 2014. *River processes: an introduction to fluvial dynamics*. New York: Routledge.

Roberts, D. 2019. These huge new wind turbines are a marvel. They're also the future. *VOX 2019*(5), 1–2.

Russell, H., G. Brooks, and D. Cummings 2011. Deglacial history of the Champlain Sea basin and implications for urbanization. In *Proceedings of the Joint annual meeting GAC-MAC-SEG-SGA, Ottawa, Ontario*, pp. 96.

Sabins, F. 1987. *Remote sensing: principles and interpretation*. New York: WH Freeman and Company.

Sahebi, M., J. Angles, and F. Bonn 2002. A comparison of multi-polarization and multi-angular approaches for estimating bare soil surface roughness from spaceborne radar data. *Canadian Journal of Remote Sensing 28*(5), 641–652.

Sahebi, M., F. Bonn, and G. Bénié 2004. Neural networks for the inversion of soil surface parameters from synthetic aperture radar satellite data. *Canadian Journal of Civil Engineering 31*(1), 95–108.

Scharien, R., R. Segal, S. Nasonova, V. Nandan, S. Howell, and C. Haas 2017. Winter Sentinel-1 backscatter as a predictor of spring Arctic sea ice melt pond fraction. *Geophysical Research Letters 44*(24), 12–262.

Shokr, M., B. Ramsay, and J. Falkingham 1996. Operational use of ERS-1 SAR images in the Canadian ice monitoring programme. *International Journal of Remote Sensing 17*(4), 667–682.

Simms, E. 2017. Nearest neighbour analysis applied to synthetic aperture radar images for the description of urban land cover and land use. *International Journal of Remote Sensing 38*(4), 1101–1113.

Singhroy, V. and R. Saint-Jean 1999. Effects of relief on the selection of Radarsat-1 incidence angle for geological applications. *Canadian Journal of Remote Sensing 25*(3), 211–217.

SOGERIVE 2006. A preferred access to the river for tourists and citizens of Longueuil. Technical report, Association SOGERIVE Inc. http://www.sogerive.com/home.html (accessed July 9, 2019).

Stantec Consulting Ltd 2016. Mactaquac Project: Final comparative environmental review (CER) report — Summary Document. Technical report, Fredericton: New Brunswick Power Corporation.

STMA 2018. Sports field dimensions. Technical report, Sports Turf Managers Association.
https://www.stma.org/knowledge_center/sportsfielddimensions/
(accessed July 9, 2019).

Strahler, A. 1952. Hypsometric (area-altitude) analysis of erosional topography. *Geological Society of America Bulletin 63*(11), 1117–1142.

Touzi, R., A. Deschamps, and G. Rother 2007. Wetland characterization using polarimetric Radarsat-2 capability. *Canadian Journal of Remote Sensing 33*(sup1), S56–S67.

Töyrä, J., A. Pietroniro, and L. Martz 2001. Multisensor hydrologic assessment of a freshwater wetland. *Remote Sensing of Environment 75*(2), 162–173.

Urquizo, N., J. Bastedo, T. Brydges, and H. Shear (Eds.) 2000. *Ecological assessment of the Boreal Shield ecozone.* Ottawa: Environment Canada.
http://publications.gc.ca/collections/collection_2014/ec/En40-600-2000-eng.pdf (accessed July 9, 2019).

Ville de Montréal 2019. *Plan métropolitain d'aménagement et de dévelopment (PMAD).* Montréal, Communauté métropolitaine de Montréal.
http://cmm.qc.ca/champsintervention/amenagement/plans/pmad/
(accessed July 9, 2019).

Walker, D. 1967. *A geography of Italy.* London: Methuen & Co Ltd.

Xia, Z.-G. and F. Henderson 1997. Understanding the relationships between radar response patterns and the bio — and geophysical parameters of urban areas. *IEEE Transactions on Geoscience and Remote Sensing 35*(1), 93–101.

Index

Printed and bound by CPI Group (UK) Ltd, Croydon, CR0 4YY

20/10/2024

01776555-0001